WITHDRAWN
L. R. COLLEGE LIBRARY

Solvents

SOLVENTS

THOMAS H. DURRANS
D.Sc.(Lond.), F.R.I.C.

Eighth Edition
revised by
ERIC H. DAVIES B.Sc.

CHAPMAN AND HALL LTD
11 NEW FETTER LANE, LONDON E.C.4.

First published 1930
Second edition 1931
Third edition 1933
Fourth edition 1938
Fifth edition 1944
Sixth edition 1950
Seventh edition 1957
Eighth edition 1971

Published by Chapman and Hall Ltd,
11 *New Fetter Lane,*
London EC4P 4EE

Set by Santype (Coldtype) Ltd Salisbury Wilts.
Printed in Great Britain
by Butler & Tanner Ltd, Frome and London

SNB 412 09680 3

Distributed in the U.S.A.
by Barnes and Noble, Inc.

Preface to the first edition

Both the number of organic solvents which are available to industry and the extent to which they are used have increased greatly during recent years, and there is in consequence a need for a scientific exposition of the technical application of solvents, more particularly in connection with the cellulose-lacquer industry.

A vast mass of valuable information germane to the subject is scattered widely throughout scientific and technical literature, but it is often presented in a profuse manner, lacking cohesion and clarity.

The author has attempted to bring this information into a concise form wherein the scientific and fundamental aspects of the subject are expressed in a readily comprehensible manner, and to show the relations of these aspects to technical usage.

The first part of this book takes the form of a more or less connected series of chapters dealing with the scientific fundamentals in a broad and simple manner. The second part is of a more utilitarian nature and deals comprehensively with individual solvents, mainly with the view of facilitating the intelligent use of these solvents in the cellulose-lacquer industry.

This book is not designed to deal with the actual manufacture of lacquers or varnishes, this aspect being the subject of a book in this series by another author.

I have to express my thanks to the Directors of Messrs A. Boake, Roberts & Co. Ltd, for the facilities they have afforded me in the compilation of this book. I have also to thank M. F. Carroll, Esq., M.Sc., and F. H. Mackenzie, Esq., for their kindness in reading and correcting my manuscript.

T.H.D.

Publishers' note on the eighth edition

This eighth edition of Solvents follows closely the pattern originally laid down by the author – this is because the book has found wide acceptance as a simple method of obtaining rapidly basic information. In the light of modern thought, however, many sections of the book have been revised radically: in particular, those sections dealing with Solvent Action, Solvent Power, and Toxicity. Fuller notes have been provided on the legal requirements associated with storing and handling solvents.

Mr E. H. Davies of Slick Brands Ltd, Croydon, has carried out the main work of revision for this new edition. Dr W. M. Morgans, Head of Surface Coatings Department of the Borough Polytechnic London, revised the material concerned with Solvent Action and Solvent Power; and Dr J. W. Daniel of the I.C.I. Industrial Hygiene Research Laboratories, was responsible for the chapter on Toxicity. We are extremely grateful to those three, and to other people and organizations who have assisted in the work.

Contents

Preface to the first edition   *page* vii
Publishers' note on the eighth edition   viii

PART I

Introduction   3
1. Solvent Action   4
2. Plasticising solvents   16
3. Solvent Balance   26
4. Viscoscity   34
5. Vapour Pressure and Evaporation Rates   41
6. Inflammability   55
7. An outline of British legal requirements   62
8. Toxicity   70

PART II

Introduction   85
Abbreviations and notes   87
1. Hydrocarbons and sundry solvents   88
2. Alcohols and their ethers   111
3. Ketones   136
4. Esters   145
5. Glycols and their ethers   166
6. Cyclohexane derivatives   179
7. Chloro-compounds   184
8. Furanes   203
9. Plasticising solvents   208
   Appendix I: Trade names   239
   Appendix II: Solubility tables   254
   Appendix III: Plasticiser proportions   261
   Index   262

PART I

Introduction

A *solvent* may be defined as a substance by means of which a solid may be brought to a liquid state. This definition, although defective and inadequate, expresses the general idea underlying the use of solvents. In the case of cellulose lacquers the object of employing a solvent is merely to provide a means of transferring a solid — cellulose ester — from one place to another in a convenient and desirable manner; when the transference has been accomplished the solvent is of no further use and has to be removed as rapidly and as completely as possible, but in a manner compatible with concomitant factors. These factors make it desirable that certain characteristics of a solvent should be known and that a solvent suitable for the purpose should be used.

The most important characteristics to be considered are:

1. Solvent power.
2. Volatility.
3. Stability.
4. Toxicity.
5. Inflammability.
6. Colour.

These characteristics are not simple phenomena, but are each composed of several factors. One and the same factor may influence more than one of the characteristics; thus, for instance, the vapour pressure of a solvent, which is largely governed by its molecular weight, affects its volatility, toxicity and inflammability; and it is thought also to be related to its solvent power and the viscosity of its solutions.

In order the better to visualize the phenomena involved in the preparation and the use of lacquers it is desirable to examine the foundations underlying them, together with their practical significance.

CHAPTER ONE

Solvent Action

Man is accustomed to 'dissolving' a wide range of materials in water and to using the solutions for purposes which have been established by long experience. However, there are fundamental differences in the nature of solutions of different materials and these have been explained only in comparatively recent years. An example is afforded by crystalline sodium chloride and non-crystalline starch, both of which dissolve in water. Solid sodium chloride can be added to water until a point is reached when further additions remain undissolved. A limit to the solubility has been reached and the solution is said to be saturated. The viscosity of such a solution varies little with concentration. In addition, the solution contains sodium and chloride ions and is an electrolyte, that is to say it will conduct an electric current. Starch, on the other hand, will continue to dissolve in water until the solution becomes very viscous and difficult to handle. The solution does not become saturated.

There is another important difference between the two. When the dilute solutions are placed in contact with a semi-permeable membrane, the sodium chloride will migrate through the membrane whereas the starch will not. Starch is classified as a 'colloid' and its solution in water as a 'colloidal solution'. This type of solution is characteristic of a large number of non-crystalline organic materials—among which are the polymers used in the surface coatings industry.

The majority of these polymers are used in organic solvents and the mechanism responsible for the 'dissolving' process has been the subject of many investigations over a number of years. One of the earliest studies was that of F. Baker[1] in 1912 on the solubility of cellulose nitrate.

Cellulose nitrate dissolves in a number of organic solvents to give solutions the viscosity of which increases with increasing concentration. There is no solubility limit but the concentrations employed are limited by conditions of application. Some varieties of cellulose nitrate are insoluble in either alcohol or ether alone yet dissolve in a mixture of the two. Baker concluded that the solvent power of the mixture arose as a result of the formation of a molecular complex between alcohol and ether. A later suggestion was of association between the cellulose nitrate and the solvent so that the true solute was a cellulose nitrate-solvent complex. Baker further suggested that the viscosity of the cellulose nitrate solution was related to the solvent power of the solvent and the most efficient solvents gave solutions of lowest viscosity.

The viscosity of a solution of cellulose nitrate in mixtures of ether and alcohol is dependent on the ratio of ether to alcohol and it was shown by Gibson & McCall[2] that this viscosity reaches a minimum at a definite ratio which is a characteristic property of a particular cellulose nitrate and does not depend on the amount in the solution. The higher the nitrogen content of the cellulose nitrate the greater is the proportion of ether required to obtain the solution of lowest viscosity.

If a solution of cellulose nitrate or acetate is cooled sufficiently a cloud develops. Assuming that the lower the temperature at which this cloud develops the higher the solvent power of the solvent, Mardles[3] concluded that solubility depends on the specific character of both the solvent and the cellulose ester.

When solvents are mixed there is usually a large divergence in solvent power (for a given polymer) from an arithmetic average of individual values. In addition, many non-solvents will become solvents if mixed together whilst in other cases mixtures of solvents produce a non-solvent blend. The solvent behaviour of mixed liquids which are known or suspected to form molecular complexes cannot be ascribed to the action of one kind of molecule or molecular complex, but, in general, wherever molecular simplification occurs in a liquid mixture there is an increase in solvent power. Liquids possessing small molecules are better solvents giving solutions of lower viscosity.Thus, in an homologous series, there is a rapid decrease in solvent power (indicated by increase in solvent viscosity) with increase in molecular weight. Mardles further concluded that, in cases where mixtures containing molecular complexes are good solvents, such as in ether-alcohol mixtures, the effect of the complex formation is masked by that due to molecular dissociation and hence simplification of the liquids.

Byron[4] found that cellulose nitrate is peptized (i.e. dissolved) by absolute alcohol at a low temperature owing to the adsorption of polymerized (associated) alcohol; such a mixture on warming becomes viscous and sets to a jelly. Adsorption decreases, in general, with rising temperature and it is known that associated liquids similarly tend to become simplified. Anhydrous ether does not peptize cellulose nitrate at any temperature and it is suggested that the function of the ether in ether-alcohol solutions of cellulose nitrate is to cause the alcohol to polymerize and thus to render it a peptizing agent for cellulose nitrate.

It was first pointed out by Highfield[5] that cellulose nitrate contains both strongly polar (hydroxyl and nitrate) and weakly polar (hydrocarbon) groups and that the most effective solvents for the polymer contain both these groups. Whilst this is true in so far as esters and nitro-hydrocarbons are good solvents for cellulose nitrate, it does not explain the insolubility of the material in alcohol. Apart from cellulose nitrate, many polymers behave in accordance with the time-honoured dictum that 'like dissolves like', that is to say highly polar solids are more soluble in solvents of high polarity than in those of low or zero polarity.

Polar and non-polar solvents

The molecules in aliphatic and aromatic hydrocarbons possess no electrical properties and, as solvents, are characterized by low dielectric constants. If, however, groups such as nitro ($-NO_2$) or hydroxyl ($-OH$) are introduced, the electron displacement in these groups confers a permanent 'dipole' on the molecules. These liquids are classed as 'polar' and are characterized by a high dielectric constant. The value of the dipole — the so-called 'dipole moment' — varies from one type of grouping to another and depends also on the size of the molecule to which the polar group is attached.

The presence of the dipole moment results in strong attractive forces between molecules, and the liquids are often described as 'associated'. This is evidenced by the fact that polar liquids possess greater viscosity than non-polar liquids of similar molecular weight. The 'assocation' also leads to higher values for certain other physical properties such as melting and boiling-points, latent heat of vaporization and cohesive energy. The latter is the internal 'cohesion' of the liquid and can be regarded as the energy required to remove a molecule completely away from the environment of its neighbours.

Cohesive energy density

This is the cohesive energy per unit volume and is related to the latent heat of vaporization by the expression

$$\text{Cohesive energy density} = \frac{\Delta H - RT}{M/D}$$

where ΔH = latent heat of vaporization
$\quad\quad$ R = gas constant (1.986) $\quad\quad\quad$ M = molecular weight
$\quad\quad$ T = absolute temperature $\quad\quad$ D = density.

It can also be defined as the energy required to separate the molecules in 1 cm³ of liquid, i.e.

$$\text{Cohesive energy density} = \frac{\Delta Hv}{V}$$

where ΔHv = molar heat of vaporization in calories.
$\quad\quad$ V = molar volume in cm³.

Solubility parameter

It was suggested by Hildebrand that the square root of the cohesive energy density, i.e. $(\Delta Hv/V)^{1/2}$, be given the name 'Solubility Parameter' and this has been assigned the symbol δ.

Solvents which possess similar solubility parameters possess similar internal cohesions. When two such solvents are brought together the two molecular species should mix freely – the solvents should be miscible in all proportions. This is true in the majority of cases provided that the solubility parameter values are less than one unit apart; the values of δ can often be used to predict the miscibility of solvents or the solubility of polymers.

However, cases arise in which materials with close solubility parameter values are not mutually soluble. This failure to mix or to dissolve spontaneously indicates abnormally high intermolecular forces in one or other of the components. For example, if two solvents A and B are brought together they will mix only if the attraction of A for A or B for B is equal to or less than that of A for B.

In a recent paper J. D. Crowley, G. S. Teague and J. W. Low[6] have shown that, in addition to cohesive energy density, hydrogen bonding and dipole moment must be taken into account. 'Hydrogen bonding' is the term applied to the bond which forms between a hydrogen atom and strongly electronegative atoms with unshared electron pairs, e.g. oxygen. It occurs in alcohols, cellulose derivatives and water. Hydrogen bonding appears to play a greater part in miscibility than does dipole moment and cases occur where it leads to miscibility or solution of a polymer in a solvent of widely differing solubility parameter.

Crowley et al. constructed three-dimensional models using as the three axes the solubility parameter, dipole moment and hydrogen bonding numbers. By the use of such a model the authors claimed to be able to predict miscibility of solvents and solubility of polymers with a greater degree of accuracy than hitherto. Using the same three parameters they were able to explain the dilution ratios for cellulose nitrate in various solvent-diluent mixtures.

Solubility parameters of amorphous polymers

Figures for the solubility parameters of solvents may be calculated from data for heats of vaporization but such a method cannot be applied to polymers. Values for these materials are determined experimentally by ascertaining the solvent or solvent mixture in which the polymer shows maximum solubility. The value assigned to the polymer is then that of the solvent or mixture used. The solubility parameter for a mixture of solvents is calculated on a volume basis.

Solubility of 'crystalline' polymers

Polymers whose molecular structure exhibits a crystal-like arrangement show abnormal behaviour towards solvents and will dissolve only if there is some interaction between polymer and solvent. PTFE and polyethylene are 'crystalline' polymers and do not react with solvents. Hence there are no solvents known at present which will dissolve these materials at normal temperature. If, however, a crystalline polymer is heated above its melting point in presence of a suitable solvent it will often dissolve but will separate when the solution is cooled.

Plasticisers are, in effect, high boiling solvents for polymers and, for effective action, the solubility parameter should be close to that of the polymer. If the polymer is of the crystalline type then, as explained above, the plasticiser will function only if there is some interaction between the two.

Cellulose nitrate solutions

If to a solution of cellulose nitrate a miscible non-solvent be gradually added, a dilution is ultimately reached, at which the cellulose nitrate begins to be thrown out of solution either as a precipitate or a gel. The solution is said to "tolerate" the addition of this definite proportion of the diluent under the conditions of temperature and cellulose-nitrate content obtaining. It is common to regard tolerances or dilution ratios as a measure of the solvent power of a solvent.

The dilution ratio is profoundly influenced by several factors, the more important of which are: (1) The nature of the cellulose nitrate or acetate. (2) The ratio of the amount of cellulose ester to that of its solvent. (3) The nature of the solvent or solvent mixture. (4) The nature of the diluent. (5) Temperature.

An empirical method for obtaining a measure of the solvent power of a solvent was originally devised by Mardles, Moses and Willstrop[7], who suggested the following conditions:

The heptane fraction of petroleum spirit is added gradually from a burette to 5 cm^3 of a 5 % weight-volume solution of a cellulose ester at 20° C. The volume of heptane required just to cause persistent turbidity after two or three minutes' shaking, divided by the volume of original solution, is the solvent power number or the dilution ratio.

In America, slightly different conditions are usually employed, and are regarded as "standard," viz.:

Two grams of "half-second" cellulose nitrate are dissolved in 20 cm^3 of the solvent under test. The volume of diluent required to cause permanent incipient precipitation at 20° C, divided by the volume of the solvent, represents the dilution ratio.

The dilution ratio can be described as the ratio of non-solvent to solvent at which the mixture ceases to be a solvent for nitro-cellulose above a given concentration.

The diluent most commonly employed for the test is toluene, the figures obtained varying widely with the diluent; thus butyl alcohol can usually be added in considerably greater quantity than can toluene, the reverse being the case with petroleum hydrocarbon (see table, p. 10).

The dilution ratio is an empirical number and represents the maximum quantity of a particular diluent which can be added to a particular solution, it represents an extreme limit and may not be of very great importance technically when considered in conjunction with other factors, such as vapour pressure or evaporation rate, since, for instance, if the diluent have an evaporation rate slower than that of the solvent it is not feasible to employ proportions even approaching that of the dilution ratio. It has been pointed out[8] that, for a technically useful result to be obtained, it is necessary that the final concentration of the solution plus diluent when precipitation occurs, should be comparable with that usually obtaining in lacquers, that is to say, from 8 to 10 % of cellulose nitrate. Thus to add, as is frequently done, two, three or four volumes of diluent to one volume of a 10 % solution of cellulose nitrate yields a solution which, at the precipitation point, bears no resemblance to the usual type of lacquer employed technically.

Cellulose nitrate dilution ratios

Solvent	Diluent			
	Toluene	Xylene	Petroleum	Butanol
Acetone	4.5	3.9	0.6	7.0
n-Amyl acetate	2.2	2.2	1.4	7.3
Isoamyl acetate	2.7	2.4	1.4	7.3
Isoamyl lactate	4.2	–	2.5	–
n-Butyl acetate	2.7	2.7	1.5	8.2
sec-Butyl acetate	2.6	2.5	1.2	8.2
n-Butyl glycollate	8.2	–	2.2	–
n-Butyl lactate	5.0	4.9	2.0	–
n-Butyl propionate	2.3	2.3	1.3	7.5
Butyl cellosolve	4.0	3.2	2.3	–
Butylene glycol diacetate	2.3	–	–	–
Cellosolve	5.0	4.7	1.0	6.9
Cellosolve acetate	2.5	2.5	1.0	7.5
Cyclohexyl acetate	2.6	–	1.2	–
Diacetone alcohol	3.1	2.9	0.5	7.8
Diamyl phthalate	2.3	–	2.0	20.0
Diamyl tartrate	9.8	–	–	–
Dibutyl oxalate	2.6	–	–	–
Dibutyl phthalate	2.8	2.7	1.7	8.0
Dibutyl tartrate	10.6	7.7	1.4	15.0
Diethyl carbonate	0.9	0.7	0.4	6.2
Diethyl oxalate	3.5	–	0.7	–
Diethyl phthalate	3.8	–	0.7	–
Diglyceryl tetra-acetate	0.9	–	–	–
Dimethyl phthalate	1.9	–	–	–
Ethyl acetate	3.4	3.3	1.0	8.4
Ethyl lactate	5.6	4.8	0.7	10.2
Furfural	2.9	–	0.15	2.1
Furfuryl alcohol	2.6	–	0.05	5.0
sec-Hexyl acetate	1.6	1.8	0.8	–
Mesityl oxide	4.4	4.1	1.0	9.2
Methyl acetate	3.0	–	0.9	–
Methyl alcohol	2.5	–	0.3	–
Methyl cellosolve	4.7	2.9	0.2	–
Methyl cellosolve acetate	2.3	1.9	0.6	–
Methylethyl ketone	4.5	3.3	–	–
1-Nitropropane	1.2	–	0.4	–
2-Nitropropane	1.3	–	0.3	–
n-Propyl acetate	3.0	2.0	1.2	–
Isopropyl acetate	2.9	2.9	1.0	8.8
Tetrahydrofurfuryl alcohol	7.8	–	0.05	12.5
Triacetin	0.9	–	–	1.0
Tributyl citrate	4.9	–	–	19.0
Tributyl phosphate	24.0	–	–	–
Tricresyl phosphate	3.3	4.2	0.7	10.5
Triethyl citrate	3.9	2.0	–	4.0

The following table shows how the dilution ratio varies with the final concentration. At high dilution the precipitation point is indefinite. The table also shows that the greater the initial concentration of the cellulose nitrate the less the amount of diluent that can be added.

Cellulose nitrate (gm)	Butyl acetate (cm³)	Toluene (cm³)	Final percentage of cellulose nitrate	Dilution ratio
1.0	20	75	1.06	3.75
2.0	20	71	2.20	3.55
3.0	20	67	3.45	3.35
4.0	20	63	4.82	3.15

Dilution ratios

The figures quoted below are the averages of determinations published by the most reliable authorities and refer in general to the results obtained with ½ sec. cellulose nitrate treated by methods approximating to that described on p. 9.

The figures should be regarded as indicative rather than accurately quantitative, as much depends on the cellulose nitrate used; for this reason figures quoted elsewhere in this book may not be in strict agreement with those in this table.

Threshold concentration

It has been suggested that a more valid comparison of solvent powers can be obtained by measuring the dilution ratio at the so-called 'Threshold Concentration' namely at just above zero concentration taken to be 0.5% w/v for practical purposes; but it has been shown that these values depend on the temperature at which the determination is made. Within the temperature range of $-10°$ C to $+50°$ C solvents having small linear molecules become more powerful as the temperature rises but those with larger linear molecules behave in the contrary manner as indicated by the following results obtained with the normal aliphatic acetates with toluene as the diluent and 0.5% w/v of cellulose nitrate.

Solvent	Mols solvent per litre at dilution ratio end point		
	$-10°$ C	$20°$ C	$50°$ C
Ethyl acetate	3.05	2.85	2.70
n-Butyl acetate	2.10	2.05	2.03
n-Amyl acetate	1.90	1.91	2.00
n-Octyl acetate	1.82	1.84	2.00

Liquids which are solvents for both cellulose nitrate and acetate generally have much higher dilution ratios with the former than with the latter, the nitrate solution being able to hold considerably more diluent; in both cases the nature of the diluent has a profound effect; indeed, quite frequently, two non-solvents, one of which is an alcohol, when mixed together give a powerful solvent mixture. A classical example is that of ether-alcohol mixture dissolving cellulose nitrate.

The following are some 'Threshold Concentration' dilution ratios with toluene and n-heptane as diluents at 20° C

Solvent	Toluene	n-Heptane
Acetone	2.57	8.10
Methylethyl ketone	2.15	6.26
Methyl n-propyl ketone	1.83	4.98
Methyl n-butyl ketone	1.64	4.15
Methyl n-amyl ketone	1.54	3.57
Methyl n-hexyl ketone	1.48	3.19
Methyl n-nonyl ketone	1.44	2.52
Dimethyl phthalate	1.69	–
Diethyl phthalate	0.91	–
Di-n-butyl phthalate	0.77	1.60
Di-n-hexyl phthalate	0.78	1.00
Di-n-octyl phthalate	0.78	0.80
Di-n-dodecyl phthalate	0.77	0.77
Methyl acetate	3.80	6.70
Ethyl acetate	2.85	5.40
n-Propyl acetate	2.30	4.58
n-Butyl acetate	2.05	4.05
n-Amyl acetate	1.91	3.62
n-Hexyl acetate	1.88	3.22
n-Heptyl acetate	1.85	3.08
n-Octyl acetate	1.84	2.88
n-Dodecyl acetate	1.85	2.46

The same phenomenon is exhibited in the case of glyceryl phthalate resins, the less highly polymerised forms of which are soluble in esters of the type of butyl acetate to which has been added an alcohol such as amyl alcohol or ethyl alcohol. Mixtures of toluene and ethyl alcohol dissolve these resins and also cellulose nitrate, whilst benzene-alcohol mixture dissolves cellulose nitrate of nitrogen content up to about 11 per cent. It is known that esters such as ethyl acetate and butyl acetate, are better solvents for mixtures of cellulose nitrate and resins when the corresponding alcohol is present than when in a pure and completely esterified state. Thus 85 to 88% ethyl acetate has a dilution ratio of 3.5 for toluene while the purer 99 to 100% has one of 3.0. The addition of quite a small quantity of butyl alcohol or benzyl alcohol to a hazy lacquer will not infrequently clarify it. More often, however, the effect of using two non-solvents simultaneously is merely that of the cumulative effects of each non-solvent separately.

Cellulose nitrate solutions

Dilution ratios are admittedly not satisfactory comparisons of solvent powers and in consequence there is an increasing tendency for research workers to employ the solubility-parameter concept coupled with hydrogen bonding values and dipole moments in the manner described by Crowley *et al* [6].

It is, however, possible to couple dilution ratios with viscosities or to make the comparisons by the latter alone using a method known as 'The Constant Viscosity Procedure' [9] which is considered to give more useful results.

The Constant Viscosity Procedure consists in determining the viscosities of solutions of plastics, such as cellulose nitrate, at three or more concentrations within the range of practical usage, e.g. 0.5 to 1.3 poises at 25° C. A viscosity-concentration curve is then plotted. A series of such curves is obtained using known mixtures of the chosen solvent and a diluent, e.g. butyl acetate and toluene; the operations are repeated with any other solvent with which it is desired to compare the chosen solvent using the same plastic and diluent. The information thus obtained is often sufficient but further curves may be constructed by plotting solvent-diluent composition against the concentration of plastic at any given chosen viscosity and related to cost or other property.

The results are in general in conformity with those given by dilution ratio measurements as regards the sequences of the solvent powers of solvents.

The procedure may be used to compare the thickening power of diluents or mixture of diluents in conjunction with any chosen solvent or mixture of solvents.

The relative solvent powers of some esters diluted with toluene or their corresponding alcohols, as determined by the Constant Viscosity Procedure, are as given under [10].

Solvent	Diluent	
	Toluene	Corresponding alcohol
Ethyl acetate	1.35	1.32
n-Butyl acetate	1.0	1.0
s-Butyl acetate	0.97	0.98
Amyl acetate (technical)	0.94	0.94

When evaluating diluents the effect of other constituents in the non-volatile part of the lacquer, e.g. resin and plasticiser, must be considered, since both the solvent balance and the solvent action are influenced by the type and concentration of these non-volatile constituents. It is therefore advisable to repeat the determinations in their presence.

A somewhat similar procedure has been used for evaluating the solvent power of hydrocarbons for resins. Here viscosity is plotted against the amount of solvent used with a given amount of resin using the various solvents which it is desired to compare. The results may be given in terms of 'Aromatic Coefficient' which is the amount of xylene compared with the amount of given solvent required to give solutions of identical viscosity with a chosen resin. Thus with an oil-modified alkyl resin the following results were obtained [10]:

Xylene (standard)	100
Ethyl benzene	106
Butyl benzene	79
Heavy naphtha	84
White spirit	54

The aromatic coefficient varies with the resin dissolved but it is independent of the viscosity at which the determination is made and of the presence of other solvents, it being an additive property.

References

[1] Baker, F., *J. Chem. Soc.*, 1912, p. 1409; *ibid*, 1930, p. 1653.
[2] Gibson & McCall, *Chem. & Ind.*, 1920, p. 172T.
[3] Mardles, *Chem. & Ind.*, 1923, p. 127T; 1924, p. 224.
[4] Byron, *J. Phys. Chem.*, 1926, 30, 1116.
[5] Highfield, *Trans. Faraday Soc.*, 1926, 22, 57.
[6] Crowley, J. D., Teague, G. S. & Low, J. W., *J. Paint Tech.*, 1966, 38, 269.

[7]   Mardles, Moses & Willstrop, *J.S.C.I.*, 1923, 127T, 207T. *Trans. Faraday Soc.*, 1933, p. 476.
[8]   Brown & Bogin, *Ind. Eng. Chem.*, 1927, p. 969.
[9]   Ware, V. W. and Teeters, W. O., *Ind. Eng. Chem.*, 1939, pp. 738, 1118.
       Ware, V. W. and Bruner, W. N., *Ind. Eng. Chem.*, 1940, p.78.
[10] Private communication from Dr. N. W. Hanson, ICI Ltd.

CHAPTER TWO

Plasticising solvents

Elasticity and plasticity

Plasticisers are substances of very low volatility which possess the power of increasing the flexibility of polymers such as cellulose esters. Flexible bodies may be of two types, the elastic type which regains its shape after the removal of the deforming force, such as rubber, and the plastic type which does not regain its shape, such as putty. Actually all bodies have both properties, elasticity and plasticity, in varying degrees, and by the appropriate choice and application of a plasticiser, it is possible more or less to control these properties in cellulose ester films, some of which are brittle and cannot be greatly deformed without fracture. Plasticisers act as lubricants and by increasing the distances between adjacent chains of polymer weaken the forces holding them together and permit greater relative movement.

Plasticisers have, in general, the effect of rendering cellulose esters less hard and more plastic. The plastic form has less tensile strength than the elastic, but by reason of its plasticity it is subject to less strain after deformation and is in consequence the more durable. A plasticised film yields to a force to a greater extent without fracture than will an unplasticised more elastic film, but the unplasticised film generally withstands the greater force; in other words, the plasticised film has the greater extensibility. In practice it is necessary to effect a compromise between this increase of extensibility or plasticity and the accompanying softness of the surface or lack of rigidity.

Evaporation of solvent and influence of plasticiser

If a solution of cellulose ester be allowed to evaporate so as to leave a film, there occurs firstly a relatively very rapid loss of solvent, the rate of loss depending, among other factors, on the vapour pressure of the solvent under the conditions which exist. This loss occurs initially, in the case of a solution made with a single solvent, at an almost constant rate until a viscosity is reached in the film which is such that the free movement of the particles of solvent to the surface of the film is restricted to a degree sufficient to retard evaporation of the solvent from the surface. At this point there follows either a rapid slowing up of the evaporation rate in the case of highly volatile solvents — i.e., low boilers — or a long drawn out retardation in the case of solvents of low volatility — i.e., high boilers. The second period is that in which the so-called 'secondary flow' occurs and to which the final gloss or smoothness of the surface is largely due. The third period, which is the one most concerned here and which critically affects the life of the film, is that in which the remaining solvent evaporates with extreme slowness, imparting an evanescent plasticity to the film. The loss of solvent which ensues during this third period in a more or less rigid unplasticised mass results in stresses caused by the inability of the mass to adjust its shape; it is probable that the cracking which occurs with old celluloid film plasticised with the volatile substance camphor is due more to the loss of plasticity than to degradation of the cellulose nitrate. In an unplasticised or insufficiently plasticised film, traces of a relatively non-volatile solvent may persist for many years, but as time elapses the amount continuously diminishes, the film in consequence becomes progressively harder and more brittle, a state of true stability never being reached. The time necessary for the film to become brittle depends on the volatility of the solvent, the thickness of the film, the temperature, the degree of exposure and similar factors.

Of these factors, the only one that is initially under control is the volatility of the solvent, but this volatility has to conform with other requirements, such as the avoidance of 'chilling' or 'blushing,' the ease of application of the lacquer, and the speed of drying and its consequent effect on 'flow.' It is obviously highly desirable to eliminate as far as possible the effect of the solvent on the ultimate nature of the film. To this end it is preferable to employ a solvent with as high a degree of volatility as is commensurate with the other requirements, in order to obtain quickly a film devoid of solvent. A film such as this would be highly brittle, and it is mainly for the purpose of avoiding or regulating this brittleness that plasticisers are used.

During the later stages of evaporation the size of the molecules of the solvent is the prime factor governing the rate of evaporation, since the larger the molecule the slower the diffusion to the surface. Hence the solvent which it is desired to eliminate from the film should have a small molecule, while for plasticisers, substances of high molecular weight are desirable, both for this reason and for the concomitant phenomenon of low volatility: there is, however, a practical limit in the case of plasticisers beyond which no useful purpose is served and which arises in the manufacture of plasticisers. Further, with increasing molecular weight there is usually a diminution of the mutual solubility between the cellulose ester and the plasticiser. Thus many of the highly complex plasticisers of excessively high molecular weight have little or no solubility in cellulose esters and are similar in their action to the resins and castor oil.

With plastic masses in contradistinction to films, the conditions which apply are somewhat different. The volatile part of the solvent evaporates as freely if not more freely from the surface of the mass than from a film: we can regard the surface of a mass as a film which is receiving further supplies of volatile solvent from the interior. There is in effect a stabilization of the surface and in consequence a higher rate of loss of solvent per given area of exposed surface can be tolerated in a mass than in a film. It is, therefore, not so essential to select solvents or plasticisers outside the 150°–250° boiling range and, indeed, the higher solvent powers of this class as compared with the plasticisers proper is of assistance where moulding under heat and pressure is concerned. With masses of considerable thickness the tendency should be to avoid the use of low boiling solvents and to rely not on loss of solvent to effect rigidity but on the adequate dispersion of the solvent in the cellulose ester. The use of a solvent plasticiser permits a plastic, which is solid at ordinary temperatures, to become on heating sufficiently soft and ductile to be pressed and moulded into desired shapes: with non-solvent softeners of the castor-oil type, the use of heat and pressure merely results in the expression of the oil without sufficient softening of the cellulose nitrate to permit moulding.

Resin plasticisers

Resins are frequently used as plasticisers for cellulose esters and they function in ways largely dependent on their chemical structure; e.g. with no solvent action on the cellulose constituent they may be regarded as an analogue of castor oil — merely held in the intercellular spaces of the cellulose but forming no 'compound' therewith. Thus, a property such as ageing resistance, even when quite free from extraneous influences, is considerably affected by chemical structure of resins. [4] Such resins give opaque mixtures, due to the existence of two solid phases, but these can be brought into a one-phase solution by the addition of a plasticiser which is a mutual solvent for both the resin and cellulose ester, thus changing the hard, elastic but brittle mass into a more plastic and more transparent mass. The second type of resin of which but few are known, is in itself a solvent for the cellulose ester. The addition of a plasticiser which is a solvent for the resin causes a lowering of the softening point: a resin which is itself a solvent for the cellulose ester requires no such addition and the softening point is in consequence not lowered. Resins which do not form solutions with cellulose nitrate lower the plasticity and increase the brittleness, while those which do form solutions increase durability and weather resistance may be used in large proportions. Plasticisers which are solvents for the cellulose ester permit the use of larger proportions of non-solvent diluents than do the non-solvent resin solvents such as those of the oil-alkyl type.

The lines of demarcation between high boiling solvents, plasticisers and resins as regards volatility cannot be accurately defined; it can, however, be said that the volatility of the resins is so low that it may be ignored, but there is great difficulty in deciding where to draw the line between the other two. The volatility of a plasticiser in a film under actual conditions of exposure is for several reasons very difficult to measure: the issue has still further been obscured by the publication of many incredible or meaningless results on the volatility of plasticisers in the free state: such results seem to be without value, firstly, because of the invalidity of the methods employed and, secondly, because they take no account of the influence of the cellulose esters and resins with which the plasticisers are commonly used.

Retention of solvent

Plasticisers also exercise the function of preventing or limiting to a large extent the retention of volatile solvent by films. Cellulose esters have the property of associating with a more or less definite amount of a solvent, forming a loosely bound complex from which the solvent evaporates much more slowly than it would if such association did not occur; with mixtures of solvents the less volatile is retained at the expense of the more volatile; plasticisers, being solvents of very low volatility, accordingly permit the low boiling solvents to leave the film with greater ease than in the absence of a plasticiser.

Resins which are not, in the absence of a common solvent, compatible with cellulose esters do not perform this function, but a resin which is itself a plasticiser for cellulose esters will prevent the protracted retention of volatile solvents.

In order that a plasticiser may be effective it must be dissolved by or associate with the solid matter of which the film is composed, since only under such conditions can a homogeneous film be obtained; a non-homogeneous film lacks mechanical strength. The association may merely be the mechanical entrapment of a limited proportion of a non-solvent plasticiser, such as castor oil in cellulose nitrate. Considering the most usual pair of solid constituents of a cellulose lacquer, cellulose nitrate and ester gum, these have no solvent action on one another and in the absence of a common solvent a non-homogeneous film of inferior strength results; frequently sufficient medium or high-boiling solvent is retained to render the mixture homogeneous for a period, but if real permanence is to be obtained it is necessary to have present a plasticiser which is a solvent for both the cellulose ester and resin. It should also be realised that the vapour pressure of the plasticiser is reduced if it is actually dissolved by the cellulose ester or other solid ingredient, and it is therefore more definitely fixed than when not in solution; furthermore, within limits, the viscosity of a lacquer solution is generally reduced by the addition of a soluble plasticiser, but may be actually increased by a substance for which the cellulose ester is not a solvent.

Critical solution temperature

If to a solid cellulose ester a solvent be added gradually, the solvent first associates with the solid causing it to gel; this phenomenon is best regarded as an absorption of the solvent by the cellulose ester, and the capacity of being thus absorbed or dissolved is an essential property in a plasticiser. If the addition of the solvent be continued, a region is reached in some cases in which two layers are formed, one layer consisting of solvent dissolved in the cellulose ester and the other of the cellulose ester in the solvent. The point at which this separation begins represents the limiting amount of plasticiser which can be used having regard merely to this consideration. If such a mixture having the two layers be heated, these layers will merge at a certain temperature, the 'critical solution temperature,' but on cooling the two layers will again form. With some mixtures the critical solution temperature is below that normally obtained in the atmosphere, in which case cooling is necessary if the phenomenon is to be observed. The phenomenon is dependent on the nature of both the solvent and the cellulose ester.

It follows therefore, that whereas some plasticisers can be used in any proportion with a cellulose ester without showing any phenomena due to lack of homogeneity, others may give rise to cloudy or opaque films or to films which exude the plasticiser (synaeresis). It is even possible to simulate the effect of a solid white pigment by using a sufficient excess of a plasticiser, such as amyl benzoate, with cellulose acetate. The opinion is widely held that solid plasticisers, such as triphenyl phosphate, are unsuitable as plasticisers, since they tend to crystallise out of the film; this is only true when the critical solution temperature is above the atmospheric temperature and the proportion of plasticiser is in excess of that which can be held in stable solution at atmospheric temperature. If this proportion be exceeded the system is unstable and the plasticiser, be it a solid or a liquid, may separate. Solids such as triphenyl phosphate under such conditions give cloudy films having a wax-like deposit on the surfaces, while liquids, such as amyl stearate and castor oil, give oily deposits. When such undesirable phenomena occur the remedy is to reduce the proportion of the particular plasticisers or to add another plasticiser, the critical solution temperature of which is well below the atmospheric.

Dilution ratio

The solvent power of plasticisers for cellulose esters or vice versa, as some prefer to regard the phenomenon, varies considerably and is of considerable significance in certain directions: if dilution ratios be regarded as measures of the solvent powers certain conclusions may be drawn. Plasticisers possessing high dilution ratios permit the use of larger proportions of non-solvent resins, such as ester gum or similar non-compatible diluent than will those having low dilution ratios: they also are the more effective in preventing cotton blush. Plasticisers have a marked influence on the gloss of a film and on the flowing properties of a cellulose ester solution, particularly on the secondary flow; this property is largely bound up with the viscosity of the solution in the later periods during drying and the viscosity in turn is largely a function of the solvent powers of the plasticiser. It follows that P plasticisers having high dilution ratios are the better for imparting gloss, for the elimination of brush marks and for the ease with which a mass may be moulded under heat and pressure.

The proportion of plasticiser which should be used in a cellulose ester film varies not only with the plasticiser but also with the effects it is desired to procure. The chief effects which are governed by the proportion of plasticiser are:

1. Plasticity.
2. Tensile strength.

3. Gloss.
4. Adhesion.
5. Elimination of volatile solvent.
6. Prevention of cotton, resin and water blushes.

The following table gives the relative dilution ratios of the more commonly used plasticisers:

Plasticiser	Dilution ratio to toluene with ½ sec. nitro cotton. HX 45 ICI
Tributyl phosphate	24.0
Dibutyl tartrate	10.65
Diamyl tartrate	9.8
Tributyl citrate	4.9
Triethyl citrate	3.9
Tricresyl phosphate	3.25
Diethyl phthalate	2.95
Butyl laevulinate	2.95
Dibutyl phthalate	2.8
Acetophenone	2.7
Dibutyl oxalate	2.6
Diamyl phthalate	2.3
Butylene glycol diacetate	2.3
Dimethyl phthalate	1.9
Benzyl laevulinate	1,6
Cyclohexyl laevulinate	1.4
Triacetin	0.9
Diglycerylether tetraacetate	0.9
Diacetin	0.35
Dicyclohexanyl adipate	
Dimethyl cyclohexanyl methyladipate	
Benzyl alcohol	
Benzyl benzoate	Little or no solvent action under the conditions of the test [1]
Cyclohexanyl phthalate	
Butyl acetyl ricinoleate	
Butyl stearate	
Castor oil	

No very definite quantitative information appears to exist on these aspects, and it is therefore only possible to make generalized approximate statements.

Small proportions of plasticisers initially increase the elasticity and tensile strength of cellulose ester films, which, in the case of plasticisers of the type of tricresyl phosphate and dibutyl phthalate, reach a maximum value with about 20 per cent. of plasticiser, [2] the plasticity and tensile strength thereafter diminish with increasing proportion; maximum gloss is also reached with similar proportions. A somewhat higher proportion is required to cause the ready elimination of residual volatile solvent, while still higher proportions favour adhesion, stability and the ease of application of the lacquer. The maximum proportions which can be used in the absence of resins are indicated in Appendix III.

Plasticised films subjected to tension exhibit elastic or reversible stretch until a certain tension is applied at which plastic or non-reversible stretch occurs. Highly plasticised films have, in general, a short region of elastic and a long region of plastic stretch before fracture occurs, whereas non- or insufficiently plasticised films exhibit a longer region of elastic stretch and little or no plastic stretch, but more force is required to produce a given amount of stretch in the non-plasticised film: colloquially it has less 'give.'

The fundamental function of a plasticiser is to increase extensibility, but while plasticisers effect this they also decrease elasticity and increase plasticity – that is, they decrease the reversible elongation or stretch and increase the non-reversible: the degree to which these two types of elongation are affected varies considerably with the plasticiser. Although no hard and fast demarcations can be made the plasticisers, in respect of this effect, may roughly be divided into three classes [2]:

1. Those which permit, to a large extent, the retention of the original elasticity of the non-plasticised film while imparting considerable plasticity or extensibility – viz., camphor, tricresyl phosphate, diphenyl diethyl urea and p-toluene sulphonamide.

2. Those which make the film highly plastic and sacrifice elasticity – viz., triacetin, amyl oxalate and tributyl phosphate.

3. Those intermediate between these extremes – e.g., the phthalates, sipalins, castor oil.

Tensile strength

The tensile strength of cellulose nitrate is affected in varying degrees by plasticisers. Van Heuckeroth [3] quotes the following figures for cellulose nitrate films plasticised with 50 per cent of various plasticisers under conditions of application similar to those obtained in factory practice. He remarks that 'tensile strength should not be stressed too much, as different conditions and varying lengths of time of ageing of the films will influence the results greatly, and in some cases may invert the order thereof.'

Plasticiser	Tensile strength
Plasticiser No. 1	581
Plasticiser No. 2	556
Diphenyl phthalate	470
Dimethyl phthalate	447
Tricresyl phosphate	382
Diphenyldiethyl urea	378
Triphenyl phosphate	363
Diethyl phthalate	348
Toluene ethyl sulphonamide	307
Castor oil	252
Diamyl phthalate	240
Dibutyl phthalate	229
Butylacetyl ricinoleate	223
Tributyl phosphate	211
Butyl stearate	203
Ethylacetyl ricinoleate	192

Viscosity

The viscosity of a plasticiser, which is a solvent for or is miscible with a polymer, determines to a considerable extent the properties of the resultant plastic; more particularly with polymers of the type of polyvinyl chloride and chloroacetate, polythene, polystyrene, polymethacrylates and butadiene-acrylonitrile.

The effect of isomerism on plasticiser efficiency has also been found of importance; increasing separation of the ester groupings tends to increase solvent power, water resistance and volatility. [6]

Plasticisers of low viscosity give the softest plastics, low in hardness and rigidity and with relatively poor electrical properties; those plasticisers whose viscosities are the least affected by change of temperature give plastics which are in general the least sensitive in these respects. On the other hand smaller proportions of the low viscosity plasticisers are required to attain particular mechanical properties and the resultant plastics are, in consequence and within limits, generally the less sensitive to temperature changes. [5] The resistance of plasticised films to fungal attack can be considerably influenced by the plasticiser present since, particulary in those cases of 'non-compounding,' the plasticiser invariably acts as a food source for fungi and micro-organisms. The piperidides of decanonic, palmitic and stearic acids and the amides of cyclic imines have all been found to be excellent anti-fungal plasticisers. [7]

References

[1] Durrans and Davidson, *Chem. and Ind.*, 1936, p. 162T.
[2] Jones and Mills, *Chem. and Ind.*, 1933, p. 251T.
[3] Van Heukeroth, Nat. Paint Varnish Lacquer Ass. Circular 485, (1935).
[4] Kraus, A., *Farbe. u Lack.*, 1964, 70, 519.
[5] $\begin{cases} \text{H. Jones, } J.S.C.I., 1948, 67, 415. \\ \text{A. Dyson, } J. Appl., Chem. 1949, p. 205. \end{cases}$
[6] Reichardt, W., *Plaste v Kautschuk.*, 1965, 12, 78.
[7] *J. Amer. Oil Chemists' Soc.*, 1964, 41, 237.

CHAPTER THREE

Solvent balance

Composition of a lacquer

A cellulose lacquer consists usually of:
1. A cellulose ester,
2. A resin,
3. A solvent of high vapour pressure,
4. A solvent of medium vapour pressure,
5. A diluent,
6. A plasticiser, and
7. A pigment or dye;

with the last item we are not here concerned.

We have to examine the general conditions which apply and the phenomena which may arise when a solution such as that indicated above, or variables of it, evaporates.

Rate of evaporation and its effects

In order to prevent the occurrence of phenomena known technically as 'chilling,' 'gum blushing,' 'cotton blushing,' and the like, it is necessary to control the order and the degree in which the various liquid components of a lacquer evaporate. There is no great difficulty in effecting such control when the cellulose ester and the resin are both soluble in the solvents employed, such as is the case with solutions of cellulose nitrate and ester gum in ethyl, butyl or amyl acetates, or in certain 'Two-type' solvents [1] such as diacetone alcohol, or ethyl lactate, which possess groups having specific solvent properties for cellulose nitrate and resins. The sole defect likely to arise in such instances is that of 'chilling' or 'water blush,' which is due to the lowering of the temperature of the film and of the air in its immediate neighbourhood to below the dew point of the atmosphere, this being caused by the excessively high rate of evaporation of a too volatile solvent. Chilling is often characterized by the early appearance of cloudiness or opacity in the still liquid film, this opacity, if not excessive, generally disappearing on warming or long standing.

Diluents

If to a solution of cellulose nitrate and resin in a single solvent there be added a diluent, it is necessary to use a diluent the rate of evaporation of which is greater than that of the solvent in order to ensure that at no instant shall the ratio of diluent to solvent be in excess of that at which either the cellulose nitrate or the resin begins to be thrown out of solution, i.e., the dilution ratio at any particular concentration shall not be exceeded. These limits can be experimentally investigated by determining the dilution ratios for a series of final concentrations (see p. 9), and it will be found that, in general, the greater the concentration of the solids the smaller the ratio of diluent to solvent permissible. If, therefore, it is desired to use the maximum proportion of diluent it is advisable to employ a mixture of solvents whose boiling-points increase progressively, e.g., ethyl, butyl and amyl acetate, but it should be borne in mind that increase in boiling-point is generally accompanied by decrease in solvent power (cf. Table of Dilution Ratios, p.10). For the same reason it is undesirable to use mixed diluents, a single diluent of suitable vapour pressure being the best, the vapour pressure of this diluent being greater than that of any of the solvents, but sufficiently low to avoid chilling.

Azeotropic mixtures

If a volatile diluent be mixed with a volatile solvent, the boiling-point of the mixture may be lower than that of either (see Chapter 4), and the ratio of diluent to solvent which evaporates may in consequence be affected. Such azeotropic mixtures, as they are termed, are of quite common occurrence, many instances being given in Part II. Azeotropic mixtures tend to cause chilling and necessitate the incorporation of a larger proportion of the solvent of higher boiling-point; further, the ratio of solvent to diluent which evaporates may be such that the solvent is removed in greater quantity than is the diluent, and precipitation may in consequence occur. The use of a diluent of considerably lower boiling-point than that of the lowest-boiling solvent will avoid or mitigate this effect. Azeotropic mixtures appear occasionally to have solvent powers different from what would be normally expected, but the phenomenon is as yet not clearly demonstrated. It would seem advisable to avoid the simultaneous use of diluents and low-boiling solvents, which give rise to azeotropic mixtures.

Solvents vary in solvent power, diluents vary in precipitating power; indeed, the demarcation between solvents and non-solvents is not clearly defined. Aliphatic hydrocarbons generally precipitate cellulose nitrate from its solutions more readily than do aromatic hydrocarbons, and these more readily than the aliphatic alcohols; the members of each class differ, more or less progressively, from one another, but the difference is not so marked. The nature of the cellulose nitrate or acetate or the resin has a profound effect; some cellulose nitrates, for instance, being soluble in ethyl alcohol while others are completely insoluble.

Cellulose ester-resin solutions

If it is desired to produce a solution of cellulose ester and resin and to use two different solvents, each of which is a solvent for only one of the solids respectively and not for the other, it becomes a matter of considerable difficulty to adjust the solvent mixture satisfactorily. This is particularly the case with cellulose acetate-resin solutions. On mixing a solution of a cellulose ester with a solution of a resin, precipitation of one or both of the solids may partially occur depending on the tolerance (see Dilution Ratios) of one solvent for the other with respect to the dissolved solids. If, however, precipitation does not occur on mixing, it may do so during the subsequent evaporation owing to the limits of toleration being exceeded, especially if the rates of evaporation of the two solvents be widely different. The effects produced by this cause are known respectively as 'cotton blush' and 'gum blush' and are characterized by the appearance of cloudiness or opacity shortly before the film sets, the opacity persisting permanently.

When it is not feasible to adjust the evaporation rates of the two solvents satisfactorily, these types of blush can be avoided by using a small proportion of a high-boiling solvent, which is a solvent for both the cellulose ester and the resins; the plasticiser can often fulfil this function.

Another phenomenon, the reason for which is not clearly understood, is the appearance of a haze when separate solutions of cellulose ester and resin in the same solvent are mixed. Instances of this are found with solutions of cellulose nitrate and ester gum in diacetone alcohol and of cellulose nitrate and kauri in ethyl lactate. In the former case the haze can generally be removed by adding a small quantity of benzyl alcohol or cyclohexanol,and in the latter case of butyl alcohol or acetate or of ethylene glycol monoethyl ether.

Water miscible solvents

Solvents which are readily miscible with water present a specific aspect. Water may be regarded merely as a non-solvent diluent such as are the hydrocarbons and the alcohols; there is no essential difference. Water has been used to replace the more expensive organic diluents in conjunction with water-miscible solvents, such as the ethylene glycol monoalkyl ethers, diacetone alcohol and ethyl lactate, but the device has no technical value on account of the powerful tendency of water to precipitate resins and cellulose esters and to produce the so-called 'orange-peel' effect. This effect, which has been likened to the pimply surface of an orange, can be caused by an excessive rate of evaporation of the solvent relatively to that of the diluent. It is frequently noticed on sprayed films when the solvent has evaporated too freely during transit from the spray-gun nozzle to the coated surface, causing a too rapid increase in viscosity. It has also been stated that by employing correct solvent/diluent blends in lacquers considerable economies in lacquer wastage may result since lower atomizing pressures and therefore less overspray can be achieved [4]. If to a lacquer in which the amount of diluent is on the border line of the maximum permissible, more diluent be added, be it hydrocarbon, alcohol or water, precipitation will occur. This is precisely what occurs when chilling takes place, the non-solvents being in excess locally, generally on the surface of the film. If the lacquer contains a sufficient excess of solvent, chilling will not occur. Where the limiting proportion of diluent is present the introduction of more diluent, in the form of water, by deposition from the atmosphere, will cause precipitation or chilling. Under such circumstances it is desirable to have a solvent which is not capable of dissolving water, as any precipitation is then limited to the surface and is transient.

Proportions of water up to about 1% increase the solvent powers of many solvents and diminish the viscosities of their solutions, hence the demand for completely anhydrous solvents is probably unsound – except, for example, with moisture-cured polyurethane lacquers where the presence of water can cause polymerization in the can.

Determination of solvent composition

In order to visualize the limiting proportions of a system of three solvent-diluent components, use may be made of triangular coordinate graphs [2]. The triangular coordinate graph system is represented by an equilateral triangle, each side of which is divided into an equal number of equal divisions, usually one hundred, so as to represent percentages. In the example shown here (curve 1) the apexes A, B, and C represent respectively 100% of three pure components such as (A) Benzene; (B) Butyl alcohol; (C) Butyl acetate. A point X along AB represents a mixture of $BX\%$ of benzene and $AX\%$ of butyl alcohol, a point Y along

AC one of *CY*% of benzene and *AY*% of butyl acetate, while *Z* along *BC* represents *CZ*% of butyl alcohol, and *BZ*% of butyl acetate. A point *P* within the triangle represents the composition of a mixture of all three, the percentage of each being given respectively by the lengths of the perpendiculars from *P* to the sides of the triangle opposite to the apex representing 100% of the particular component, thus *Pa* represents the percentage of benzene, *Pb* that of butyl alcohol, and *Pc* that of butyl acetate, the sum of *Pa*, *Pb* and *Pc* being one hundred. The ratio of benzene to butyl acetate is found by joining *B* and *P* and extending the line to cut the line *AC* at *b'*. In a similar manner the ratio of benzene to butyl alcohol and of butyl alcohol to butyl acetate can be found.

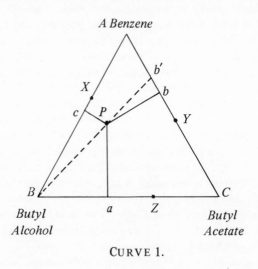

CURVE 1.

A practical application of the method may here be given. Suppose it is desired to determine the solubility limits of a given percentage of cellulose nitrate in mixtures of butyl acetate, butyl alcohol and benzene. A series of determinations is made of the ratio of butyl acetate to diluent (butyl alcohol, benzene and mixtures of these). Thus the point *A* (curve 2) represents the composition of a mixture of butyl acetate and benzene that will just dissolve the given percentage of cellulose nitrate; the point *B* the equivalent mixture of butyl acetate and butyl alcohol, while *C*, *D*, and *E* represent the composition of mixtures of all three. Any mixture the composition of which is represented by a point below the line *AEDCB* will not dissolve the given percentage of cellulose nitrate, whilst any mixture whose composition falls into the area above the line will give complete solution. The knowledge thus obtained allows other factors to be correlated, for instance, that of cost or rate of evaporation. Thus, although benzene is cheaper than butyl alcohol it has greater precipitating power, and a larger proportion of butyl acetate is therefore required than if some of the benzene had been replaced by the more costly diluent butyl alcohol; the cheapest mixture may, for instance, be represented by the point *E*.

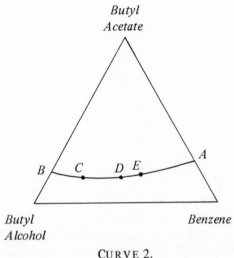

CURVE 2.

Triangular coordinates can similarly be used to visualize the change in composition of a three-component solvent mixture on evaporation, and to determine if at any time during the evaporation the composition could be such that precipitation of the cellulose nitrate could occur. Other uses to which the system may be put are the determination of solvent compositions for mixtures of a cellulose ester and a resin, for the variation of viscosity [3] and of cost.

As a further illustration of the practical application of triangular coordinates, let it be supposed that it is desired to produce the cheapest possible lacquer having 10% of a certain quality of cellulose nitrate dissolved in a mixture of 'low boiler,' 'medium boiler' and diluent; the lacquer to have a definite viscosity and blush resistance. Let the apex *D*(curve 3) represent 100% of diluent, *L* represent 100% of 'low boiler' and *M* 100% of 'medium boiler.' Determine the dilution ratio of the 'low boiler' and diluent for a mixture that will just dissolve the 10% of cellulose nitrate, let X be the point on the line DL which represents this mixture, similarly determine the point Y for the 'medium boiler' and diluent. Within certain limits of accuracy a straight line XY represents all mixtures of the solvents and diluent which will just dissolve 10% of the cellulose nitrate. In general, XY will not be a straight line, but will be curved towards D. If greater accuracy is required the curve of the

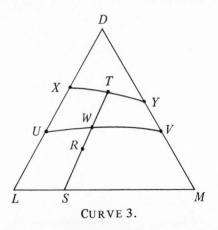

CURVE 3.

line may be found by determining the dilution ratios for known mixtures of the two solvents with the diluent. The area DXY now represents mixtures which will not dissolve the requisite proportion of cellulose nitrate and such mixtures must therefore be left out of further consideration.

Next consider one of the other properties which it is wished to establish in the lacquer, the resistance to chilling for instance. Having decided what are the limiting conditions of humidity, temperature and rate of drying that the lacquer must withstand without chilling, determine by systematic trial and error tests what ratio of 'low boiler' and 'medium boiler' just gives signs of chilling under the prescribed conditions and obtain the point S on the line LM which represents this ratio; repeat the tests with mixtures of all three liquids, the compositions of which lie on the line XY, and obtain the point T. Join S and T. Here again, ST is not necessarily a straight line, depending largely on the possible formation of low-boiling azeotropic mixtures; the determination of points such as R may be advisable. The region $XLST$ represents solutions which will chill under the given conditions, and can therefore be eliminated from further consideration. Consider now a third property, viscosity for example. Determine by any of the methods the composition of the mixture of 'low boiler' and diluent which gives the desired viscosity with 10% of the cellulose nitrate and obtain the point U on the line LD; similarly determine the point V, using the 'medium boiler' and diluent. Join U and V. The curvature of the line UV may be determined by actual test with mixtures, the composition of which are represented by points on the line ST, for instance W. Should either U or V fall above X or Y respectively it may still be possible to produce a lacquer having the desired properties, since the lines UV and XY may cross. It may quite well happen that the desired viscosity cannot be obtained with mixtures of the 'low boiler' and diluent without precipitation occurring. Should this be so it will be necessary to make viscosity determinations with mixtures lying close to the line XY, but considerable inaccuracy may arise, as there may be somewhat abrupt changes of viscosity as the insoluble region is approached. Having determined the viscosity line UWV, any third property, such as cost or 'flow,' can be determined along the line VW. It is easy, for instance, to calculate the cost of lacquers represented by points along the line WV, the point W generally representing the cheapest mixture of solvents. Flow may be measured as the viscosity after a certain proportion of the solvent has evaporated, and similarly determined along VW.

The subsequent incorporation of resin and plasticiser into the lacquer will have some effect and may necessitate a slight adjustment (see page 19, 50).

Certain short cuts may be made; thus the determination of the line XY is not absolutely necessary, and the point W may be found without determining S and T, but the limitations of the lacquer cannot then be visualized so well.

References

[1] Keys, *Ind. Eng. Chem.*, 1925, p. 1120.
[2] Hoffman, Reid and Stoppel, *Ind. Eng. Chem.*, 1928, p. 431.
[3] Cochrane and Leeper, *Chem. and Ind.*, 1927, p. 118T.
[4] Meades, K. G., *Ind. Finishing*, 1965, **41**, (8) 66.

CHAPTER FOUR
Viscosity

Viscosity may be broadly defined as 'resistance to flow,' and is of considerable technical importance in connection with lacquers.

The viscosity of a lacquer may be adjusted in several ways; thus different grades of cellulose nitrates may be employed, these being available in various viscosities, or the viscosity may be varied by suitable choice of solvent or solvent mixture. Viscosity and vapour pressure are apparently to a considerable extent concomitant properties, the greater the vapour pressure of a solvent the lower the viscosity of its solutions.

The following table [1] shows the percentage of dry half-sec. cellulose nitrate which can be dissolved in various solvents to give solutions having a viscosity of 10 secs (Cochius):

Acetone	15
Ethyl acetate	12
Amyl acetate	9
Butyl propionate	8
Propyl butyrate	7.5
Diacetone alcohol	4
Cyclohexyl acetate	4

Effect of non-solvents

The viscosity of a solution may in general and within limits be increased by the substitution of part of the solvent by a non-solvent, but inasmuch as the admixture of a non-solvent with a solvent sometimes gives a mixture having increased solvent powers, no hard and fast rule can be formulated; for this reason the partial substitution of a non-solvent for a solvent sometimes initially results in a decrease of viscosity, whilst the continued replacement of the solvent beyond a certain proportion causes an increase, and finally gelation or precipitation of the cellulose nitrate.

This phenomenon of the diminution of viscosity is marked with ethyl cellulose as shown in the following table relating to 10% solutions and their viscosities at 20° C.

Solvent A and viscosity of solution	Solvent B and viscosity of solution	Proportion A:B giving min. viscosity	Viscosity of mixture
n-Butanol 160	Xylene ∞	30:70	42
Methylated Spirit 35	Xylene ∞	25:75	16.25
sec. Butanol 216	Xylene ∞	30:70	29.8

With a given cellulose nitrate the viscosity of its solution in ether-alcohol mixture is dependent on the ratio of ether to alcohol [2], and reaches a minimum at a definite ratio. Indeed, the viscosity of a solution of cellulose nitrate is fundamentally bound up with the nature of the solvent, but in general it may be said that the greater the solvent power of a liquid the lower the viscosity of its solutions [3]. The same may be said of solutions of cellulose acetate. Thus with mixtures of acetone and alcohol the minimum viscosity is obtained when the ratio of alcohol to acetone is about one to six [4], with benzene and acetone, when the ratio is one to five; a mixture containing 25% of a 3:1 blend of isobutanol : toluene gives solutions of the lowest viscosity [16].

Viscosity determinations are of considerable value for the purpose of selecting and standardising the cellulose ester or ether which is to be used to prepare a lacquer and also for standardising the resultant lacquer. A lacquer destined for use in a spray gun should initially be less viscous than a brushing lacquer, and in both cases it is of importance to have a knowledge of the progressive variation in viscosity consequent upon the partial evaporation of the solvents.

The viscosity of a cellulose-ester solution does not increase proportionately with the concentration of the solution except at extreme dilutions; with concentrated solutions a small increase in the content of cellulose ester greatly increases the viscosity [6], between concentrations of 10% and 20% the viscosity increases geometrically [5]. It follows, therefore, that in order to adjust the viscosity of all but very thin solutions, it is better to vary the proportion of cellulose ester than to alter the composition of the solvent, since the latter practice may cause some undesirable effect, such as 'blushing,' or bad 'flow.'

When it is desired to reduce the viscosity of a solution by means of 'thinners,' these should have the same composition as that of the liquids in the original mixture: if, however, a lacquer has become viscous by reason of the evaporation of part of the solvent, it is then necessary to use thinners having a larger proportion of the low-boiling constituents, and in order to use the correct amount of these it is best to analyse the lacquer first and then to make the necessary additions.

Viscosity changes during evaporation

The viscosity of cellulose-ester solutions, the solvents of which have not been allowed to evaporate, becomes less with age, but subsequent agitation of the solution causes the viscosity to return ultimately to its original value [7]. There is usually a limited rapid decrease in viscosity, attributed to the rupture of weak linkages; the phenomenon is not found with highly nitrated cellulose, with regenerated nitro-cellulose or in the presence of water. The decrease is followed by a slower continuous fall attributed to the hydrolysis of glucoside links [15]. The presence of basic substances accelerates the reduction of viscosity with age; sunlight and heat have the same effect, but water and urea appear to have a stabilizing effect.

It is possible to obtain considerable control over the viscosity or flow of a lacquer during all stages of its drying. In general, it may be postulated that solvents of low boiling-point yield solutions of low viscosity, and that increasing boiling-point is accompanied by increasing viscosity. This is particularly noticeable in a homologous series such as methyl, ethyl, butyl and amyl acetates. The phenomenon may be ascribed to the change in molecular concentration which must accompany the change in the size of the molecule; thus there is a larger number of molecules of ethyl acetate in a given weight than in the same weight of amyl acetate. The molecular proportion of ethyl acetate is therefore the larger in a solution containing a given percentage of cellulose nitrate.

As has been pointed out above, a small increase in the cellulose-ester content of a lacquer greatly increases the viscosity. It follows from this that when a solvent evaporates from a solution of cellulose ester dissolved in a single solvent the increase of viscosity is relatively slow at first, but continues to increase at an accelerating rate up to a point determined by the concomitant slowing up of the rate of evaporation of the solvent; thus the rate at which the viscosity increases is at first comparatively slow, becoming more rapid, and then slower and slower. By employing a mixture of low and medium boilers, the rate of evaporation during the period in which the increase of viscosity is rapid may be slowed and the rapid increase of viscosity slowed in consequence. This is a desirable effect to obtain, since lacquers which increase in viscosity too rapidly flow badly, and are difficult to manipulate. It is necessary, however, to employ some quantity of low-boiling solvent since 'medium boilers' and 'high boilers' do not lead to a sufficiently rapid increase of viscosity after application; the flow is in consequence too protracted and the film develops waves or otherwise varies in thickness.

Functions of solvents and diluents

The diluent may affect the changes in viscosity considerably. The viscosity of a solution may, in general, and within limits, be increased by substituting diluent for solvent; but since the addition of a diluent may occasionally increase the solvent power of a solvent, the viscosity sometimes initially decreases on adding a diluent, while a further quantity of diluent results in increased viscosity, and finally in precipitation of the cellulose ester. The converse effects may be produced when a low-boiling diluent evaporates from a lacquer, the solvents of which are medium or high- 'boilers,' namely, an initial increase of viscosity followed by a diminution. In general, if the 'low boiler' and diluent evaporate at the same rate the viscosity of the remainder will increase, but it is possible to maintain a more or less constant viscosity over a short period of evaporation by using a diluent which evaporates more rapidly than the 'low boiler.'

The change of viscosity in the later period of the drying of a film may be slowed by employing high-boiling solvents, this is commonly done with brushing lacquers designed for amateur use, the prolongation of the semi-fluid state allowing ample time for manipulation. Such lacquers tend, however, to develop waves or other unsightly defects besides yielding films which lack permanency. It is better to rely on medium boilers and plasticisers to obtain delayed drying.

The viscosity during the final stages of drying depends largely on the proportion of plasticiser and its plasticising power as contrasted with its power of imparting elasticity. High gloss, which is to a large extent the result of prolonged secondary flow (see p. 21) can be obtained by plasticisers having high plasticising power.

The functions of the various classes of solvents may be summarized as follows:

(1) *Low boilers* of high solvent power to enable concentrated solutions of low viscosity to be prepared and to obtain quick initial drying and increase of viscosity.

(2) *Medium boilers* to check the rate of evaporation and thus to reduce defects caused by excessive rates and also to impart good flowing properties.

(3) *High boilers* to protract the period of drying for long periods when necessary, and to impart good brushing properties and high gloss.

(4) *Plasticisers* to knit all the solid constituents into a homogeneous whole, to impart flexibility, gloss and permanence, and to reduce brittleness.

(5) *Diluents* to reduce cost, to control viscosity, and, occasionally, to dissolve resins.

Resin solutions

The viscosity of solutions of resins follows on quite different lines to those of the cellulose esters. The viscosity of solutions of the former is practically constant with solvents of a homologous series, such as benzene, toluene and xylene, or di-, tri-, and perchloroethylene [8]. The viscosity depends markedly on the constitution of the solvents, being much greater with saturated than with unsaturated substances. The presence of a hydroxyl group in the solvent appears to increase the viscosity. Drummond [9] divides synthetic resins into two classes—those soluble in 'polar' solvents, such as alcohols or ketones, which contain hydroxyl or carbonyl groups; and those soluble in 'non-polar' solvents, such as hydrocarbons. The former class includes phenolic resins, acrolein resins and glycerylphthalate resins, whilst in the latter class are cumarone, albertol, oil-soluble phenolic resins, and naphthalene-formaldehyde resins.

The measurement of the viscosity of lacquers is obviously of considerable technical importance, and much attention has been given to the methods and instruments which may be employed. The instrument of greatest utility for lacquers is known as the 'falling sphere viscometer' [10]. The method of its operation has been exactly standardized by the British Standards Institution [11], and consists, briefly, in measuring the time of fall of a clean steel ball between two points in a vertical glass tube filled with the solution, the temperature of which has been carefully adjusted. For details of the test the reader is referred to BS 188:1957. The results are given in poises, (kinematic viscosity) and can be converted to stokes by multiplying by the density of the liquid (dynamic viscosity).

The following figures have been quoted [12] for the comparative viscosities of 15% solutions of 4 sec NC in various solvents:

	seconds
Acetone	20
Methyl acetate	42
Ethyl acetate	92
Amyl acetate	127
Isobutyl acetate	131
Butyl glycol	200
Methyl glycol	225
Cyclohexanone 299	
Methyl cyclohexanone	304
Ethyl lactate	397
Butoxyl	426
Diacetone alcohol	610
Butyl glycollate	625

Another authority [13] gives:

	Viscosity of 10% solution at 25°C
Acetone	270
Methylethyl ketone	503
Methyl acetate	540
Dioxane	1295
Methyl cellosolve	1960
Methyl cellosolve acetate	2310
Ethyl lactate	2730
Diacetone alcohol	4095

while for 8% solutions of cellulose acetate (54.5% CH_3COOH) we have [14]

Acetone	46
Methylethyl ketone	155
Ethyl acetate	101
Dioxane	130
Methyl cellosolve	163
Diacetone alcohol	very viscous

References

[1] Hiag, *Holzverkohlungsindustrie.* Konstanz.
[2] Gibson and McCall, *Chem. and Ind.,* 1920, p. 172T. *Cf.* Mardles, *J. Chem. Soc.,* 1924, p. 2245.

[3] *Cf.* Section on 'Solvent Power' and *Ind. Eng. Chem.*, 1939, p. 738; 1940, p. 78.

[4] *Brit. Chem. Abs.*, 1928, B519.

[5] Merz, *Farb. Ztg.*, 1922, **34**, No. 44, 2567.

[6] *Ind. Eng. Chem.*, 1928, p. 195.

[7] *Brit. Chem. Abs.*, 1926, p. A677.

[8] *Brit. Chem. Abs.*, 1926, p. B248.

[9] *J. Inst. Rubber Ind.*, 1928, p. 44.

[10] Gibson and Jacobs, *J. Chem. Soc.*, 1920, p. T473.

[11] British Standards Institution, 2 Park Street, London W.1.

[12] *Nitrocellulose*. Berline, 1937, Vol. 1, p. 7.

[13] Hoffman and Reid, *Ind. Eng. Chem.*, **21**, p. 955.

[14] C. Bogin, *Paint, Oil and Chem. Rev.*, 1942.

[15] H. Campbell and P. Johnson, *J. Polymer Sci.* 1950, **5**, 443.

[16] 'Isobutanol in Surface Coatings.' ICI Technical Publication, 1968.

Vapour pressure and Evaporation Rates

It is not an exaggeration to say that all substances give off a vapour and therefore exert a vapour pressure. The vapour pressures of substances vary to a wide extent, and that of most solids is quite inappreciable at ordinary temperaturs to all but the most delicate instruments. The vapour pressure of most liquids at ordinary temperatures is confined to quite a limited range, viz., pressures up to one atmosphere, but the variation within this range of the vapour pressure of solvents is of paramount importance, since it is one of the prime factors governing the rate at which evaporation takes place.

Control of vapour pressure

It is necessary to be able to control, in some degree, the rate at which the liquid medium of a lacquer evaporates in order:

1. To prevent the deposition of water from the atmosphere.
2. To control the 'flow' of the lacquer.
3. To prevent the premature precipitation of any ingredient which might cause the film to be non-homogeneous.
4. To prevent undue contraction of the film.

For any given single solvent the rate of evaporation is governed by several factors:

1. The vapour pressure of the liquid at the temperature under consideration.
2. The rate at which heat is supplied.
3. The conductivity for heat of the liquid.
4. The specific heat of the liquid.
5. The latent heat of evaporation of the liquid.
6. The degree of association of the molecules.
7. The surface tension of the liquid.
8. The molecular weight of the liquid.
9. The humidity of the atmosphere.
10. The rate at which the vapour adjacent to the liquid is removed.
11. The vapour density of the solvent.
12. The solvent power of the solvent.

With mixtures the rate of evaporation is further governed by:

1. The molecular attraction of one component by another.
2. The depression of the vapour pressure of one component by another.
3. The viscosity, if the solute be a colloid.

The problem is obviously one of considerable complexity. It may be possible at some future date accurately to evaluate the influence of all these factors and thus to control the manner in which a lacquer shall dry, but at present it is only possible to form a rough judgment, and it is necessary to allow a large margin of safety in order to accommodate conditions over which it may not be possible to have control.

Many essential data regarding the various factors enumerated above are missing, and it is as yet very difficult to apply technically such information as is available.

The first attempt in this direction was to classify solvents as low, medium and high 'boilers,' according to their boiling-points, and although the classification was admittedly unsatisfactory it permitted some degree of control to be obtained. It was customary to consider as 'low boilers' those solvents with boiling-point below 100° C, as 'medium boilers' those boiling from 100° C up to an indefinite temperature above the boiling-point of amyl acetate, e.g., about 150° C, whilst those of still higher boiling-points were termed 'high boilers,' these last merging into the 'plasticiser' class, which included substances like benzyl alcohol.

Boiling-points and vapour pressures

It was soon realized that boiling-points represent the temperatures at which the vapour pressures of liquids are the same as the atmospheric pressure and do not accurately correspond to their vapour pressures at the temperatures at which lacquers are ordinarily used, namely, atmospheric or slightly higher. Thus the examination of vapour pressure-temperature curves shows that in some instances the curves relating to two substances actually cross and that the relative vapour pressures vary with the temperature irrespectively of the boiling-point. This phenomenon is generally occasioned by the molecules of one of the liquids being 'associated' at normal temperatures, i.e., two or more molecules are loosely combined; this association causes a diminution in the vapour pressure of the substance, but it breaks down progressively as the temperature rises. The phenomenon is observed particularly among those substances having hydroxyl groups, as in the case of alcohols and acids. It is for this reason that alcohols of quite low boiling-point, such as propyl alcohol, evaporate more slowly than some of their esters of higher boiling-point, such as propyl acetate. Another instructive example is that of the closely similar substances cyclohexanone and cyclohexanol, these boil at nearly the same temperature $(150^\circ \text{ C} - 160^\circ \text{ C})$, but cyclohexanone evaporates ten times as rapidly as cyclohexanol.

Not only may a single liquid form associated molecules, but liquids on mixing sometimes associate one with another. In the latter case the simplest relations have been found to hold.

Vapour pressure and mixed liquids

The vapour pressure of two liquids, A and B, which are miscible in all proportions and do not react chemically with one another or associate, is given by the expression:

$$100P_{A+B} = M_A P_A + (100 - M_A)P_B$$

where M_A is the molecular percentage of liquid A,

P_B and P_A the respective vapour pressures of A and B at the given temperature, and

P_{A+B} the vapour pressure of the mixture at the same temperature.

If intermolecular attraction between the two liquids exists, the mixtures have abnormal vapour pressures, either higher or lower than the values given by the above equation. Thus in the case of acetone and chloroform the gradual addition of one of them to the other causes a diminution of the vapour pressure until a minimum is reached, and then an increase until the vapour pressure of the one in excess is reached. The same phenomenon occurs with methyl acetate and chloroform, but it is not common with organic solvents. It is more usually found that the gradual addition of one solvent to another causes the vapour pressure progressively to increase until the maximum is reached, and thereafter to diminish. The mixture having the maximum vapour pressure (minimum boiling-point) is termed an azeotropic mixture of minimum boiling-point, and has, of course, a lower boiling-point than either of its two components. This mixture of minimum boiling-point is not necessarily of the same percentage composition as that having the maximum vapour pressure at ordinary temperature [1]; thus the composition of mixtures of minimum boiling-point of benzene and alcohol varies with the pressure, as shown in the following table [2]:

Pressure (mm Hg)	b.p. of azeotropic mixture (° C)	Percentage alcohol
760	68.25	32.4
570	60.0	28
380	49.9	26
241	35.0	23.3
200	34.8	22

If these figures are extrapolated to normal temperature the propor-tion of alcohol is found to be about 20%.

It should be noted that the boiling-point of the azeotropic mixture is lower than that of either component at the same pressure.

If the vapour pressure of mixtures of the two liquids A and B be

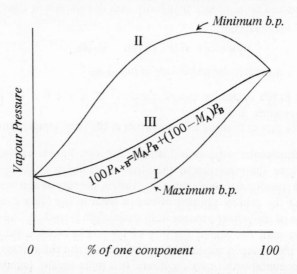

plotted against the composition of the mixture, it is found that the curve takes one of three forms, as shown in the diagram above. Form III is that normally shown by non-associated liquids such as mixtures of ethyl acetate with propyl alcohol; it is practically never a straight line. Mixtures of maximum boiling-point, such as acetone with chloroform, give a curve of Form I, whilst mixtures of minimum boiling-point, commonly found, such as ethyl acetate with ethyl alcohol, give Form II. Other types of curve are theoretically possible, but only one minimum or maximum can occur with substances that are miscible in all proportions [3].

A number of instances are known in which three liquids form a mixture having a lower boiling point than any of its components, e.g., ethyl alcohol, chloroform, hexane; isopropyl alcohol, ethyl acetate, and cyclohexane. In all probability the number of components may be still greater but no actual observations appear to have been recorded.

Azeotropic mixtures

The reason for the scarcity of azeotropic mixtures having more than two components seems to lie in the probability that for a system of three or more components each component must be capable of forming a binary azeotrope with at least one of the others separately, so that with an increase of but one in the number of components the chance of the formation of an azeotropic mixture comprising all the components is greatly lessened.

It has been shown [12] that the composition of the vapour arising from a mixture of liquids evaporating in free air may be profoundly influenced by minor modifications in the conditions and that the composition of the vapour under certain conditions may be different to that of the azeotrope at the same temperature.

In actual practice azeotropic mixtures have effect in several ways. The increase of vapour pressure which accompanies the formation of an azeotrope has a distinct influence on the flash-point of a mixture. The increased vapour pressure causes a greater quantity of solvent to be evaporated into the air space of a container at any given temperature than if no azeotrope were formed, and since the flash-point is largely governed by the proportion of inflammable vapour, the necessary proportion is reached at a lower temperature. Thus [13] 25% of butanol flash p. 95° F (35° C) when mixed with 75% of xylene flash p. 74° F (23° C) gives a mixture flash p. 68° F (20° C). Similarly the proportion for maximum explosivity (cf. p. 59) is reached at a lower temperature. The increase in vapour pressure will also cause the rate of evaporation to be greater and may result in 'chilling' or other defects due to an excessive rate of evaporation; furthermore, the deposition of water during chilling may give rise to ternary azeotropes which have generally still higher rates of evaporation. It is also possible that the formation of an azeotrope may affect the solvent power of a mixture.

The formation of an azeotrope affects the relative rates at which the constituents of a mixture evaporate, since the azeotrope tends to evaporate as though it were a single substance. It is advisable, therefore, when using a non-solvent diluent to choose the solvent so that either no azeotrope is formed or if it be formed that it shall be one having a higher proportion of non-solvent than solvent, in order that the non-solvent may be removed more rapidly than the solvent; otherwise cotton blush or gum blush may occur if the concentration of these solids be too near to the saturation point of the mixture.

Molecular weight and vapour pressure

If no azeotrope is formed by two miscible liquids, the vapour pressure of the mixture is directly proportional to the relative amounts of the two liquids and their respective vapour pressures at the temperature under consideration, namely the normal atmospheric temperature. By adding a medium or high boiler to a low boiler we slow up its rate of evaporation more or less proportionally to the amount added. Theoretically a medium boiler will reduce the vapour pressure and consequently slow up the rate of evaporation of a low boiler more than will a similar weight of a high boiler. A substance of low molecular wieght is more effective than a similar mass of one of high molecular weight, since it is the number of molecules that matters, and mass for mass the substance of low molecular weight has the larger number of molecules.

If a layer of lacquer be allowed to evaporate, the rate at which the solvent leaves the solution slows progressively, the lacquer becomes more and more viscous, and the molecules of solvents diffuse more and more slowly to the surface from which evaporation takes place. At this stage the size of the solvent molecule is probably the prime factor governing the rate of evaporation, but vapour pressure still has some effect. Generally, large molecules and low vapour pressure are concomitant phenomena; substances the molecules of which are large evaporate more slowly than those having small molecules.

When a liquid evaporates it gives up energy and its temperature drops. In order that it may continue to evaporate, it is necessary to supply heat from an external source, otherwise the temperature would drop until evaporation ceased. The drop in temperature is governed by three factors, the specific heat and the latent heat of evaporation of the liquid, and its conductivity for heat; of these factors the latent heat is the most important.

Latent heat of evaporation

The latent heat of evaporation of a liquid can be found approximately by Trouton's formula.

Trouton found, in 1884 [4], that the amounts of heat required to evaporate quantities of different liquids taken in the ratio of their molecular weights are approximately proportional to their absolute boiling-points; in other words, the product of the molecular weight and the latent heat of evaporation is proportional to the absolute boiling-point, or expressed symbolically:

$$M \cdot H = cT,$$

where c is a constant (the value of which for normal liquids is 20 to 21, such liquids being hydrocarbons, esters, ethers and ketones, whilst for associated liquids such as alcohols and water the value is higher, being about 26), M is the molecular weight of the liquid, T the absolute temperature ($^\circ$ Kelvin), and H the latent heat of evaporation.

Methods have been devised for obtaining a measure of the blush resistance of solvents. One method is somewhat similar to that for measuring the solvent powers of solvents by determining their dilution ratios, and consists in diluting the solvent with blush-inducing materials until a faint blush can be obtained under standardized conditions [7].

Blush numbers

The blush numbers found for certain solvents are as follows:

Ethyl acetate, 85%	0
Ethylene glycol ethyl ether	0
Butyl acetate	1.05
Butyl propionate	1.75
Pentexel	2.00
Ethylene glycol butyl ether	3.90
Ethylene glycol ethyl ether acetate	4.00
Ethyl lactate	4.00

Another method [11] consists in estimating the degree of blushing of 10% solutions of cellulose nitrate under varying degrees of humidity.

Lowering of temperature on evaporation

Empirical evaluations of the lowering of the temperature of solvents on evaporation have been made by Gardner and Parks [5] under more or less standardized but arbitrary conditions; the results, however, seem to be of somewhat doubtful validity.

The lowering of temperature of a solvent evaporating in free air causes the air in its immediate neighbourhood to fall in temperature, and if the humidity of this air by sufficiently great a temperature below that of the dew-point may be reached and water will be deposited. The following table [15] gives some interesting results obtained with the wet-bulb thermometer.

Solvent	Wet bulb depression ($^\circ$C)	b.p. ($^\circ$C)
Acetone	30.5°	56.3°
Methylethyl ketone	21.5°	79.6°
Ethyl acetate	21.5°	77.2°
Methylpropyl ketone	12.0°	102.0°
n-Butyl acetate	6.5°	126.5°
Methylbutyl ketone	6.5°	127.0°
Methylamyl ketone	3.0°	150.0°

Chilling

If the solvent contains a substance insoluble in water, for instance, cellulose nitrate or resin, this deposition of water may cause their precipitation with the consequent formation of a white opaque film on the surface of the solution; this effect is known as chilling, or blushing. If the solvent is one able to dissolve water to a considerable extent, such as ethyl lactate, diacetone alcohol, or ethyl alcohol, this chilling effect may be delayed, since the water may be dispersed throughout a greater depth of the solution. If, on the other hand, water is completely insoluble in the solvent no chilling can occur, since the water will merely deposit on the surface and not penetrate at all. It follows that a hygroscopic solvent is more likely to give rise to chilling than is a non-hygroscopic one (cf. p. 47).

It is frequently found that a lacquer, normally satisfactory, will chill in abnormally damp weather; by determining the dew-point or humidity of the atmosphere it is possible to select a solvent composition suitable for the occasion; the greater the humidity the greater the proportion of high or medium boiler. The determination of the dew-point should be a regular factory practice.

Chilling is not infrequently prevented by raising the temperature either of the lacquer or the air to a degree sufficiently high to prevent the dew-point being reached on subsequent evaporation; another method is to dry the air either by chemical means, by compression, or by condensing out the water by cooling and reheating the dry air again before use. The compression method is the most convenient when the lacquer is to be applied by a spray-gun. Chilling can also be eliminated by diminishing the rate of evaporation, this being effected by enclosing the lacquered article in a more or less stagnant atmosphere partially charged with the vapours of the solvent.

Chilling which has taken place may disappear at a later period during the drying of a lacquer owing to the evaporation of the water and the re-solution of the precipitated solid in the high-boiling solvent remaining. The addition of a solvent of high molecular weight, and hence of high boiling-point, is usually an effective method of eliminating a tendency to chilling in a lacquer, since the high boiler not only redissolves precipitated solids but also, as already shown, slows down the rate of evaporation of the low-boiling solvent to which the chilling is initially due.

'Sand-papering' and blushing

A defect which sometimes accompanies the use of a spray-gun is that of a rough surface known as 'sand-papering' or 'pimpling'; this is caused by the passage of too large a volume of air, caused commonly by the use of excessive pressures. The large quantity of air promotes such rapid evaporation of the low-boiling solvents that the lacquer reaches the 'work' in a semi-solid or plastic state and is unable to flow sufficiently; the remedy is to use either a lower air pressure or partly to replace the more volatile constituents with less volatile.

Many of these difficulties may be avoided and several advantages gained by heating the lacquer and air. At temperatures around 70° C, only about one half of the air pressure is required as compared with normal temperature spraying, on account of the lower viscosity of the lacquer, with the further advantage that the proportions of low-boiling solvents and diluents can be diminished and the proportion of solids increased; this results in thicker films being deposited for a given quantity of spray with consequent reduction in the number of coats, less strain on the solvent recovery plant, less spray fog and less risk of fire and health hazard [14].

Cellulose lacquers usually contain both a cellulose ester and a resin, the latter being added for the purpose of imparting 'body,' gloss, hardness or adhesiveness to the resultant film. It is thus necessary that the solution should at all periods during its evaporation contain a sufficiency of solvent or solvents to keep in solution both the cellulose ester and the resin. Should the resin solvent by reason of its higher rate of evaporation fall below a certain proportion, the resin will begin to precipitate, the film will become cloudy and the defect known as

	Evaporation	Boiling Ranges ($^{\circ}$C)
Ethyl ether	1	34–35°
Methylene chloride	1.8	40–42°
Acetone	2.1	55–56°
Methyl acetate	2.2	56–62°
Ethyl acetate, 98/100%	2.9	74–77°
Benzene	3	80–81°
Benzine	3.5	67–100°
Trichlorethylene	3.8	86–88°
Ethylene chloride	4.1	81–87°
Isopropyl acetate	4.2	84–93°
Toluene	6.1	109.5–110°
Propyl acetate	6.1	97–101°
Methanol	6.3	64–65°
Perchlorethylene	11.0	119–122°
Propyl alcohol	11.1	96–98°
Butyl acetate, 98/100%	11.8	121–127°
Butyl acetate, 85%	12.5	110–132°
Monochlorobenzene	12.5	130–131°
Xylene	13.5	137–139°
Diethyl carbonate	14	120–130°
Isopropyl alcohol	21	79.5–81.5°
Isobutyl alcohol	24	104–107°
Butanol	33	114–118°
Glycol monomethyl ether	34.5	115–130°
Glycol methyl ether acetate	35	138–152°
Glycol monoethyl ether	43	126–138°
Glycol ethyl ether acetate	52	149–160°
Dichlorobenzene	57	167–180°
Amyl alcohol	62	129°
Ethylbutyl carbonate	68	135–175°
Cyclohexyl acetate	77	170–177°
Ethyl lactate	80	155°
Diacetone alcohol	147	150–165°
Butyl glycol	163	164–182°
Benzyl acetate	393	213–216°
Butyl lactate	443	170–195°
Ethylacetyl glycolate	464	181–195°
Benzyl alcohol	1,767	204–208°
Glycol	2,625	191–200°

'gum-blush' will arise. In a similar manner the cellulose ester may be precipitated and give rise to 'cotton-blush'.

It is thus desirable that the solvent of lowest vapour pressure should be a solvent for both the cellulose ester and the resin; frequently the plasticiser will perform this function effectively; indeed, it is a desirable property in a plasticiser that it should be a solvent for all the solid constituents.

The diluent used in any lacquer, being usually a non-solvent for either the cellulose ester or the resin or for both, should obviously evaporate at an early stage in the process, and therefore it should preferably be of the low-boiling type.

Evaporation rates

The rates of evaporation given on p. 50 were determined by placing $0.5cm^3$ of solvent on to filter paper and measuring the time required for complete evaporation. The time required for the evaporation of ethyl ether is taken as unity, and the figures given for the other solvents represent the time each requires for evaporation compared with that of ethyl ether; the figures are therefore relative only [6].

Another authority [7] gives the following:

Substance	Evaporation Period
Ethyl ether	1
Methyl acetate	1.4
Acetone	2.1
Ethyl acetate	5
Methylethyl ketone	6.3
Ethyl propionate	9
n-Propylacetate	12
Isobutyl acetate	15
Butyl acetate 85%	24
Isobutyl propionate	26
Propyl propionate	39
Amyl formate	40
Butyl propionate	60
Amyl acetate	63
Propyl butyrate	73
Ethyl butyrate	80
Isobutyl butyrate	92
Amyl propionate	135
Butyl butyrate	160
Diacetone alcohol	317
Amyl butyrate	443
Cyclohexyl acetate	491
Methyl cyclohexyl acetate	554

The agreement with the previous table is obviously not good, variations between the results of different investigators being excessively wide.

Very little reliance, however, can be placed on most of the published information on evaporation rates as the phenomenon is greatly affected by several factors which are difficult to control or standardize. Moreover, the affinity of the solvent for the dissolved solid with which it is used has a profound effect in technical practice.

Evaporation from a thin film of plasticiser (0.1 g tritolyl phosphate, 0.9 g solvent) has been used as a test method with results more akin to those experienced under practical conditions [8]. Butyl acetate is taken as unity and other solvents related thereto by simple calculation so that liquids with numbers greater than unity are faster evaporating and those less than unity are slower evaporating,. Thus, acetone, on this scale, has a figure of 10.2 whilst methyl cyclohexanone has a figure of 0.18.

The apparatus used is known as the ESTL Evaporometer and it was developed from the Rudd-Tysall Evaporometer [9].

It is interesting to study the investigations made by Hoffmann who has determined the rates for several solvents by using a modified form of the apparatus devised by Poleich and Fritz, reasonably consistent results being obtained. These rates are based on that of butyl acetate, which is taken as 100. It has been confirmed that results can be calculated to a reasonable accuracy from the formula: (vapour pressure $\times$ molecular weight) +11.

The experimental results are given as rates of evaporation at 25° C, viz.:

$$\frac{\text{time of evaporation of n-butyl acetate} \times 100}{\text{time of evaporation of other liquid}}$$

The two sets of results are given below.

Liquid	Experimental results	Calculated results
Acetone	850	975
Ethyl Acetate	485	582
Alcohol 95%	193	183
Toluene	186	186
s-Butyl acetate	162	180
s-Butyl alcohol	110	87.5
n-Butyl acetate	100	100
s-Amyl acetate	73	85
Xylene	65.5	62
n-Butyl alcohol	45.7	31.7
Glycol monoethyl ether	37	37.6

The order of this table is in good relative agreement with that given on p. 51, where the results are given as evaporation times.

By experimental means, the existence at 20 to 25 deg. C of a number of constant evaporating mixtures has been established and it is interesting to compare these mixtures with those of constant boiling-point where known. The effect due to the molecular association of the alcohols is shown in the lowering of the proportion of alcohol at the lower temperature. Methylethyl ketone also seems to be strongly associated at normal temperatures, and hence evaporates more slowly than would be expected from a consideration of its boiling-point.

	Composition % by weight at	
Mixtures	20–25 deg. C	Boiling point
Benzene	46.5	–
Ethyl acetate	53.5	–
Benzene	71.5	60
Methyl alcohol	28.5	40
Carbon tetrachloride	85.5	84
Ethyl alcohol	14.5	16
Carbon tetrachloride	82.5	71
Methylethyl ketone	17.5	29
Toluene	46.5	32
Ethyl alcohol	53.5	68
Toluene	50	31
Isopropyl alcohol	50	69
Toluene	79	–
s-Butyl alcohol	21	–
Toluene	93	–
s-Amyl alcohol	7	–
Xylene	66	–
s-Amyl alcohol	34	–
Xylene	36	–
s-Butyl alcohol	64	–
Xylene	71	–
n-Butyl alcohol	29	–
Xylene	82	–
Glycol monoethyl ether	18	–

References

[1] *Cf.* Wade and Merriman, *J. Chem. Soc.,* 1911, p. 997.
[2] Ryland, *J. Amer. Chem. Soc.,* 1899, p. 384, *Cf. Z. physikal. Chem.,* **47,** 445.
[3] Marshall, *J. Chem. Soc.,* 1906, p. 1350.
[4] Independently discovered also by Pictet in 1876 and by Ramsay in 1877.
[5] *U.S. Paint Manuf. Assoc. Circular,* No. 236, 1925.
[6] *Oil and Colour Td. J.,* 1930, p. 560.
[7] Sharples Chemicals Inc., N.Y., U.S.A.
[8] Shell Chemical Co.
[9] Rudd and Tysall, *J.O.C.C.A.* 1950, **33,** 520.
[10] *Brennstoff-Chem.,* 1924, **5,** 371.
[11] *Cf.* De Heen, *J. Chem. Phys.,* 1913, **11,** 205.
[12] U.S. Industrial Chemical Co., Inc., N.Y.
[13] Lewis and Squires, *Ind. Eng. Chem.,* 1937, **9,** 109.
[14] W. A. Chitson, *Paint Technology,* 1943, p. 36.
[15] *Cf. Products Finishing,* 1951, May, p. 38.
[16] Park and Hoffmann, *Ind. Eng. Chem.,* 1932, **24,** 132.

CHAPTER SIX

Inflammability

Flash-point and boiling point

The inflammability of a lacquer is governed initially by the flash-point of its volatile ingredients. Practically all lacquers contain either solvents or diluents of the low-boiling class which flash at ordinary temperatures, e.g., acetone, alcohol, ethyl acetate, benzene, toluene and ligroin.

The flash-point of a mixture of inflammable liquids is not necessarily identical with that of its lowest-flashing constituent — it may be higher or lower. Mixtures of inflammable vapour and air are explosive or inflammable within quite limited ranges of concentration; a mixture may contain too little or too much of the inflammable vapour for ignition to take place on the application of a flame. The flash-point of a liquid is the temperature at which this minimum concentration is reached.

If to a liquid of given flash-point we add another of higher flash-point, we should, in general, expect the flash-point of the mixture to lie between those of the two liquids taken separately, but it is frequently found that the mixture has a flash-point lower than that of any of its constituents. This phenomenon is caused by the formation of an azeotropic mixture of lower boiling point than those of its constituents. Under such conditions a concentration of vapour sufficiently high to support ignition is reached at a lower temperature than that required by any of the constituents.

The flash-point of a lacquer may be raised in several ways. The first and obvious way is to eliminate wholly or in part those constituents of low flash-point, this generally entails eliminating the low boilers. A second way is to choose low boilers which do not form mixtures of low boiling-point. A third way is to increase the proportion of high boilers at the expense of the medium boilers. Fourthly, low-boiling non-associated liquids may be replaced by associated liquids of similar boiling-point e.g., alcohol for benzene. Last and best, a small proportion of highly volatile non-inflammable liquid [1] such as methylene chloride, ethylene chloride, trichloroethylene, perchloromethane, 1,1,1,-trichloroethane, or carbon tetrachloride may be used.

British Railways classify inflammable liquids in two classes: (1) Those flashing below 73° F (22.8° C) (closed test); (2) those flashing between 73° and 150° F (22.8-65.6° C); both classes are considered to be dangerously inflammable.

A useful empirical relation has been found, by Ormandy and Craven, between the flash-point of hydrocarbons and the initial boiling-point, viz.:

$$\text{Flash-point (° C)} = (0.736 \times \text{boiling-point}) - 72.$$

This relation is only accurate when the hydrocarbon is free from traces of impurities more volatile than the hydrocarbon itself.

Ignition temperature

The fire risk attending a lacquer is determined more by its ignition temperature than by its flash-point, i.e., the temperature at which the lacquer will *continue* to burn after ignition; this temperature is generally higher than the flash-point and it determined by the conditions obtaining, viz., a sufficient supply of oxygen; the heat evolved during combustion; the rate at which the heat of combustion is dissipated; and the manner in which the lacquer is disposed; a porous substance, for instance, will act as a wick and promote combustion. The susceptibility of a lacquer to continuous burning can be governed to a slight extent by the inclusion of a substance which leaves a non-inflammable residue on ignition, such as triphenyl phosphate. This device is more effective with the solid film left by a lacquer than with the lacquer itself, but it should be borne in mind that cellulose nitrate is merely a modified form of gun-cotton and contains in itself sufficient oxygen to ensure its complete combustion in a closed space in the absence of air. The only way to extinguish a burning film of cellulose nitrate is to reduce its temperature to below its ignition point; for instance, by flooding with water.

Explosion and explosive limits

The use of inflammable volatile liquids also involves the risk of explosion, and the conditions under which such explosions may occur have been investigated to a very considerable extent. If the temperature of an inflammable volatile liquid be gradually increased, the air above the liquid becomes progressively richer in inflammable vapour; at a certain concentration the air-vapour mixture can just be ignited on the application of a flame or spark of sufficient thermal intensity. This concentration, which corresponds with the flash-point, is termed the lower explosive limit. As the temperature, and hence the concentration, increases, the ease with which the mixture can be ignited also increases and the combustion becomes progressively more violent until a maximum explosivity is reached. On still further increasing the concentration of the vapour the violence of the explosion gradually diminishes until finally the mixture will no longer support combustion, the upper explosive limit having been reached.

In the case of hydrocarbons the explosive limits can be correlated with the nature and composition of the liquid [2]. At the upper limit there is twice as much inflammable vapour as there is in a mixture which contains just enough air to correspond with complete combustion; at the lower limit ignition fails to take place when the mixture contains twice this quantity of air. It appears that these limits are controlled by the amount of heat liberated by the combustion; the heat of combustion at the limits being just insufficient to maintain the mixture at the ignition temperature.

The upper explosive limit of a substance in a homologous series of which the first terms are known can be calculated from a typical equation. Thus, if n be the number of oxygen atoms required theoretically for the combustion of one molecule of a substance, then for the lower limit the number required is $2n - 1$ atoms for a hydrocarbon and $(3n - 2)$ for methyl or ethyl alcohol.

The temperature of the air-vapour mixture affects the explosive limits, increase in temperature causing an increase in the range [4].

Temperature ($^\circ$C)		Lower limit
50°	1.5	4.85
300°	1.3	6.2

The figures refer to toluene-air mixtures, ignited by an electric spark in a glass apparatus.

Large changes in pressure also affect the limits [5].

Le Chatelier [6] found that for a mixture of substances whose lower limits taken separately contain N_1, N_2, N_3, etc., percentage of each substance in air, the composition of the lower limit mixture is given by

$$n_1/N_1 + n_2/N_2 + n_3/N_3 + \ldots = 1,$$

where n_1, n_2, n_3, etc., are the percentages of the different substances in the mixture.

The figures given in this Table are considered to be reliable:

Explosive limits	Per cent. by volume in air		Flash points	
	Lower limit	Upper limit	°C	°F
Acetone	2.15	13.0	−16.7	2
n-Amyl acetate*	1.1	–	–	–
n-Amyl alcohol*	1.2	–	57	134
Amyl chloride	1.4	–	3	38
Benzene	1.4	8.0	−11	12
n-Butyl acetate*	1.7	15.0	39	102
n-Butyl alcohol*	1.7	–	47	116
Isobutyl alcohol	1.68	–	22	72
Carbon disulphide	1.0	50.0	−22	−8
Cellosolve*	2.6	15.7	40	104
Cellosolve acetate*	1.71	–	51	124
Cyclohexane	1.31	8.35	3	37
1,2-Dichlorethane	6.2	15.9	17	63
1,2-Dichlorethylene	9.7	12.8	–	–
Diethyl ether	1.7	48.0	−41	−40
Ethyl acetate	2.18	11.5	−4	25
Ethyl alcohol	3.28	19.0	14	57
Ethyl formate	3.5	16.5	−19	−2
Furfural	2.1	–	56	133
Furfuryl alcohol	1.8	16.3	75	167
n-Heptane	1.0	6.0	–	–
n-Hexane	1.25	6.9	–	–
Methyl acetate	4.1	13.9	−13	9
Methyl alcohol	5.5	21.0	0	32
Methyl n-butyl ketone	1.22	8.0	23	73
Methyl cyclohexanone*	1.15	–	55	130
Methylethyl ketone	1.81	11.5	−7	19
Naphtha. V. M. & P.	1.2	6.0	–	–
Paraldehyde	1.3	–	27	81
n-Propyl acetate	2.0	–	14	57
Isopropyl acetate	2.0	–	8	46
n-Propyl alcohol	2.5	–	22	72
Isopropyl alcohol	2.5	–	12	53
Tetrahydrofurfuryl alcohol	1.5	9.7	80	176
Toluene	1.27	7.0	4	40
Turpentine*	0.8	–	32	90
o-Xylene	1.0	5.3	30	85

With the substances marked * explosive concentrations cannot be reached at temperatures below 25° C (77° F) provided no unevaporated spray remains in the air.

Heat of combustion

Richardson and Sutton [7] have investigated experimentally the explosive properties of the vapours of certain lacquer solvents at normal temperatures under conditions which approximate to those obtaining in spray booths. They point out that in order that an explosion may occur sufficient heat must be liberated by the burning vapour to heat the adjacent layers to their ignition temperature, and that the thermal conductivity of the vapour must be high enough to permit this necessary flow of heat; when the amount of the solvent vapour in the air is low, the thermal conductivity of the mixture is nearly the same as that of air, and the heat of combustion is the chief factor governing the explosion under these conditions. i.e., at the lower explosive limit; the higher the heat of combustion the greater the explosivity.

The following are the heats of combustion in calories per gram of some volatile liquids in common use:

Heptane	11,375	Ethyl ether	8,807
Hexane	10,636	Acetone	7,370
Paraffin	10,340	Alcohol	7,130
Toluene	10,150	Ethyl acetate	6,103
Benzene	9,960	Methyl alcohol	5,331
Tetralin	10,150	Methyl acetate	5,371
Dichloroethane	2,720	Isopropyl alcohol	7,970
		Butyl alcohol	8.626

At first sight it appears anomalous to state that heptane has a higher explosivity than ether, but a necessary condition is that the proportion of the vapour in the air is sufficiently high. It follows, given the necessary conditions, that mixtures of heptane and air explode with greater violence than corresponding mixtures of ether and air.

It will be noticed that the heats of combustion for alcohols and esters are, in general, much lower than those for hydrocarbons; a greater concentration of the former is, therefore, necessary if an explosive mixture is to be formed.

Liquids having boiling-points above about 113° C do not form explosive mixtures with air at 25° C, since a sufficiently high concentration of the vapour cannot be obtained. A point of practical interest is that concentrations which can be inhaled without producing pronounced narcotic effects are nonexplosive, but in any given space the possibility of the existence of local concentrations above the lower explosive limit should not be overlooked.

SOLVENTS

Auto-ignition Temperatures [3]

	°F	°C
Acetone	1118	604
n-Amyl acetate	714	379
Isoamyl acetate	715	379
n-Amyl alcohol	621	327
Isoamyl alcohol	667	353
Benzene	1076	580
Benzyl acetate	862	461
Benzyl alcohol	817	436
n-Butyl acetate	790	421
n-Butyl alcohol	693	367
Isobutyl alcohol	825	441
t-Butyl alcohol	901	483
Butyl carbitol	442	228
Butyl cellosolve	472	244
n-Butyl propionate	800	427
Camphor	871	466
Carbon disulphide	257	125
Castor oil	840	449
Cellosolve	460	238
Cellosolve acetate	715	379
Cyclohexanone* [3]	1032	557
Cymene	921	494
Decahydronaphthalene	504	262
1,2-Dichlorethane	775	413
sym-Dichlorethyl ether	696	369
Diethyl ketone*	1125	608
Diethylene glycol	444	229
Ether	366	186
Ethyl acetate	907	486
Ethyl alcohol	799	426
Ethyl benzene*	1026	553
Ethylene glycol	775	413
Ethylene glycol diacetate*	1154	635
n-Heptane	452	233
n-Hexane	477	247
Methyl acetate	935	502
Methyl alcohol	887	475
Methyl cellosolve	551	288
Methyl cyclohexanone*	1102	595
Methylene chloride	1224	662
Methylhexyl ketone*	1060	572
Paraldehyde*	1005	541
Isopropyl acetate	860	460
n-Propyl alcohol	812	433
Isopropyl alcohol	852	456
Isopropyl ether	830	443
Toluene	1026	552
Turpentine	488	253
Xylene	920	493

INFLAMMABILITY

Auto-ignition

A mixture of inflammable vapour and air may ignite without the actual application of a flame or a spark, provided its temperature be sufficiently high; this temperature is termed the self- or auto-ignition temperature. This temperature is known in Europe as the spontaneous ignition temperature and a close relationship is felt to exist between this property and the drop ignition temperature [8] so that a form of classification has been drawn up [9].

Thompson [10] determined the auto-ignition temperatures of a number of lacquer solvents. His results are, however, profoundly affected by the material of which the apparatus for determining the temperatures is constructed, and must be taken with due reserve.

There seems to be no periodic relationship connecting the auto-ignition temperature and specific gravity, vapour pressure, heat of combustion, oxygen content or size of the molecule [11].

References

[1] Brit. Pat. 243031; 286724; 302390. U.S. Pat. 1793726.
[2] Thornton, *Phil. Mag.*, 1917, **33**, 190.
[3] *Cf.* Associated Factory Mutual Fire Insurance Co., Boston, Mass., U.S.A. *Ind. Eng. Chem.*, 1940, p. 881. Except substances marked *.
[4] Author's determinations.
[5] *Chem. and Met. Eng.*, 1937, p. 733.
[6] *Compt. rend.*, 1898, **126**, 1344.
[7] *Ind. Eng. Chem.*, 1928, p. 187.
[8] ASTM D 286–30.
[9] German Electrical Engineers Society 0173/v. 43.
[10] *Ind. Eng. Chem.*, 1929, p. 134. *Cf.* Moore, *Chem. and Ind.*, 1917, p. 110.
[11] *Ind. Eng. Chem.*, 1927, p. 1335; Masson and Hamilton, *Ind. Eng. Chem.*, 1928, p. 813.
   See also "Memorandum on the Use and Storage of Cellulose Solutions," Form 826, and Statutory Rules and Orders, 1934, No. 990, H.M. Stationery Office, London, W.C.2.

CHAPTER SEVEN

An outline of British legal requirements

Handling and storing solvents or products containing solvents needs some basic understanding of the legal requirements involved; it may be reckoned that all legislation is aimed at minimizing any hazards associated with solvents and most of the inherent minor risks can be eliminated by common sense coupled with some legal knowledge. Manufacturers of every type of solvent employ experts on storage and handling whose help is readily available to customers, and combining such advice with one's personal knowledge of a particular set-up and its legal requirements should minimize any problems encountered.

Internationally, requirements concerning labelling, packaging, handling, and transportation are being examined by a UN committee and it may well be that from this work some degree of standardization in these spheres will eventually emerge.

Locally, bye-laws enacted by the relevant authority must also receive close attention since such bye-laws may lay down special requirements applicable to the vicinity in which the solvents are manipulated. Frank discussion with interested officers is undoubtedly well worth while since their advice and help will always be readily given. As an illustration of the scope involved, a list of interested persons embraces factories inspector, buildings inspector, public health inspector, factory doctor, fire chief, petroleum inspector, and insurance company inspector.

'Inflammable' and 'Inflammability'

'Inflammable' and 'Inflammability' are words increasingly criticized in modern technical parlance, mainly because the use of the prefix *IN* (which invariably means *not*) seems illogical and references to 'flammable' and 'flammability' are becoming more and more popular. Unfortunately, it is not easy to change legal terminology and this means that 'inflammable' is still employed as the all-important word in most current definitions, on labels, and on the majority of commercial documents. Efforts to rationalize the situation are in train [1] which aim at calling all liquids with a flashpoint below 90° F 'flammable', but no statutory steps have so far been taken. In view of this, 'inflammable' has been retained throughout this book.

The 1961 Factories Act

The 1961 Factories Act does not specifically mention solvents as such, but it must be clearly understood that certain sections can apply in their intent to processes handling solvents or products containing solvents. Section 4 deals with ventilation, section 18 with vessels containing dangerous liquids, section 30 with precautions against dangerous fumes, and section 63 with removal of fumes from place of work; means of escape in case of fire is also relevant, and in the event that circumstances in a particular plant were such as to make enforcement of any of the preceding doubtful, then the factories inspector might well invoke the Electricity Regulations, although these would apply mainly to fire risks.

Two additional Orders [2] [3] are still in force and refer to special requirements. Inflammable solvents are mentioned in the first but no definition is given and the second is rather more exacting since it lays down specific requirements for such things as ventilation, location of plant, medical examination of workers, washing facilities, etc., these being necessary where benzene, carbon tetrachloride, sulphur chloride, trichlorethylene or any carbon-chloride solvent is manipulated. In addition, some control is exercised over the length of period of employment in manipulating carbon disulphide; benzene is less rigidly controlled and toluene, xylene etc, are not mentioned.

The Petroleum Act

The Petroleum Act [4] has made considerable impact upon British industry although in its original conception this legislation was aimed at the storage and handling of motor fuels. The reason for this impact is, of course, the ever increasing use of solvents derived from petroleum sources and it is therefore imperative to understand clearly certain definitions laid down by the Act:

'PETROLEUM' includes crude petroleum, oil made from petroleum
                    or from coal, shale, peat or other bituminous
                    substances, and other products of petroleum.

'PETROLEUM SPIRIT' means such spirit as, when tested in the manner set forth in Part 2 of the Second Schedule to this Act, gives off an inflammable vapour at a temperature of less than 73° F (22.8° C)

'PETROLEUM, MIXTURE' means any mixture of any other substance with petroleum which when tested in the manner set forth in Part 2 of the Second Schedule to this Act, gives off an inflammable vapour at a temperature of less than 73° F (22.8° C)

There are other definitions included in the Act (as well as Orders in Council made under the Act [4]), but for purposes of inflammability the above three are the most important.

The practical difficulties of keeping within the meaning of the Act are enhanced by the formation of azeotropic mixtures.

In its simplest form the problem may be described as follows. Suppose solvent A (petroleum but *not* petroleum spirit) be mixed with solvent B, which can have any origin so that the *blend* flashes below 73° F (22.8° C), then that *blend* constitutes a Petroleum Mixture; the petroleum part itself does not *necessarily* have to be a petroleum spirit. This basically simple phenomenon is the one which causes many flash-point problems in using and handling surface coatings, polishes, cleaners, thinners, etc. and it must be remembered that these problems apply both to the manufacturer and to the user of the product. Confusion is confounded when one examines the list of products and solvents [5] which are classified as being within or without the Act. About the only simple advice one can give is the platitude 'better be safe than sorry', that is to say, err on the side of the Act rather than away from it.

The enforcement of the Petroleum Act is in the hands of major Local Authorities (in London, the Greater London Council, elsewhere the County Councils) by Petroleum Officers and requires that users should hold a petroleum licence to handle petroleum spirit or mixture from the relevant Authority; a nominal fee is payable for this licence. The licence will lay down conditions of storage, maximum quantity permitted, site of storage etc. and the original 1928 Act contains a scale of penalties for not complying with the Act or with any of the conditions laid down in a particular licence. Requirements as to labelling, departments, storage areas or tanks are included and transported goods must be clearly labelled:

PETROLEUM $\left.\begin{array}{c} \text{MIXTURE} \\ \text{SPIRIT} \end{array}\right\}$ GIVING OFF AN INFLAMMABLE HEAVY VAPOUR NOT TO BE EXPOSED NEAR A NAKED FLAME

together with the name and address of the consignee. [6,7,8,9]

Other controls are covered by the various Orders in Council so that there are virtually no loopholes allowing non-compliance with the law.

The part of the Petroleum Act which interests practically every British user of inflammable liquids is, of course, the Second Schedule 'Test apparatus to be used and manner of testing Petroleum therewith so as to ascertain the temperature at which it will give off Inflammable Vapour.' The apparatus is specified in accurate engineering detail and the finished article is known as the Abel apparatus [10]: exact requirements are laid down for carrying out the test and the routine described must be followed if acceptably reproducible figures are wanted. It is the view in some quarters that one operator can obtain variations of between 2 and 3 deg F with the same apparatus and solvent, whilst different operators may well achieve as much as 5–6 deg F difference; obviously a margin of safety should therefore always be allowed. The Abel apparatus is capable of giving closed-cup results in the range $66°$ F – $120°$ F ($18.9°$ C – $48.9°$ C) flash-point. The American test apparatus is known as the Tag-closed Tester [11] and again test operation is specified. For materials with flash-point higher than this the Pensky-Marten [12] apparatus is generally used and quoted.

The Cellulose Solutions Regulations

The next most important legislation involving solvents are The Cellulose Solutions Regulations [13], a complex and difficult document. The object is to establish precautions for the handling, use, and storage of 'cellulose' solutions and once again it is imperative to possess a clear understanding of the definitions involved.

'Cellulose solution' means a solution of any derivative of cellulose in however small an amount (and regardless of the presence of any other material) in an 'inflammable liquid'.

'Inflammable Liquid' means any liquid used or intended to be used with a cellulose solution possessing a flash-point (by the Abel apparatus) of 90° F (32.2° C) or less. There is no limitation as to the *type* of liquid — it may be a single liquid or a mixture of liquids and its/ their origin is immaterial — the criterion is 90° F (32.2° C) flash with a cellulose derivative association. Obviously this higher flash-point makes the Regulations more stringent since the scope is greater. But one is forced to ask the question: suppose a non-cellulose derivative petroleum-free solution is to be manipulated, then, provided it flashes above 73°F (22.8°C), what restrictions can be applied to such manipulation? This anomalous state of affairs is another reason for the proposed modified legislation already mentioned [1] but at present the Cellulose Solutions Regulations are British law and as such must be adhered to. No licence is needed under these regulations but the requirements concerning ventilation, fire exits, fireproofness, general safety, and electrical safety are quite stringent.

The Petroleum Act applies to cellulose solutions if the liquid used fulfils the definition of 'petroleum mixture' and the flash-point of the final lacquer is below 73° F (22.8° C) and it is this fact which causes much confusion; the labelling 'Petroleum Mixture etc' is necessary on account of this 'petroleum' content but *not* because a cellulose solution is involved — formulate a cellulose solution with a flash-point above 73° F (22.8° C) and it need carry no 'petroleum mixture' label. (Cellulose solutions which flash above 90° F (32.2° C) and are therefore outside the scope of cellulose solutions regulations are invariably known as High-Flash Lacquers.) The one requirement about labelling in the Cellulose Solutions Regulations is that which calls for a declaration on the can if more than 15% by weight of benzene (C_6H_6) is present; no person under 16 years of age is allowed to manipulate a product requiring this label.

Ministry of Transport Rules

The Petroleum Act, or Additional Orders, lay down specific requirements for harbour or canal authorities to make proper bye-laws controlling the transport by water of petroleum spirit but these requirements do not apply to other products and reference must be made to the Ministry of Transport Rules [14] for guidance in this very wide sphere; it should be realized that these Rules have no statutory authority but are generally accepted as if they had — this usually appearing in the shipping contract drawn up between sender and carrier.

Two sections are relevant:

(which are also inflammable.)	'poisonous substances' and this specifically lists the following: carbon tetrachloride, methylene chloride, pentachlorethane, tetrachloroethylene, chloroform, methyl bromide, o-dichlorobenzene, nitrobenzene.
Section 5	which covers inflammable substances but groups them as: A flash-point below 73° F (22.8° C) and B flash point between 73° F (22.8° C) and 150° F, (65.5° C) and then further subdivides into $A1$ not completely immiscible with water and M miscible with water. The method of test for flash-point is Abel.

Internationally accepted labels are now used for affixing to packages which contain such products and the Rules give colour plates of these. A fairly exhaustive list of products is included in the Rules and this embraces a large number of solvents and products likely to contain solvents. Type of packing, minimum ullage and position on vessel are all considered and although the Rules are extensive they represent a realistic approach to a complex problem.

Transporting 'Dangerous Goods' by air freight (i.e. not passenger machines) is subject to quite sensible control and the definition of Dangerous Goods covers a wide but reasonable range of products which, of course, includes solvents under the description 'goods likely to endanger the safety of the aircraft or the persons on board'. Requirements for packing, maximum capacity, ullage, etc. are laid down together with the labelling necessary and the information which must be given about the goods, (e.g. a basic outline of the constitution in the case of lacquers etc.). Flammable liquids are designated as those having an Abel closed cup flash-point of 73° F (22.8° C) and corrosive liquids are those which cause severe damage to living tissue. Poisonous substances, Class B, are those which are dangerous when taken internally or by external contact, (e.g. chlorinated hydrocarbons). A further enactment embracing solvents which directly affects industry is that relating to the payment of Customs and Excise duty on light hydrocarbons, etc., oils [15] (in the region of several shillings per Imp. Gallon for many commonly used hydrocarbon solvents). The definitions are again interesting: 'Hydrocarbon oil means petroleum oils, coal tar and oils produced from coal, shale or peat or other bituminous substance and all liquid hydrocarbons, but does not include such hydrocarbons or bituminous or asphaltic substances as are solid at a temperature of 60° F (15.5° C) or gaseous at a temperature of 60° F (15.5° C) and under a pressure of 1 atmosphere'.

'Light oils' means hydrocarbon oils of which not less than 50% by volume distils at a temperature not exceeding 185° C or of which not less than 95% by volume distils at a temperature not exceeding 240° C or which give off an inflammable vapour at a temperature of less than 73° F (22.8° C) when tested in the manner prescribed by the Acts relating to petroleum. In addition, there is a special definition of material to be classed as heavy oil [16] 'of which not more than 50% by volume distils at 170° C and not more than 90% at 210° C.'

The whole question of duty payment goes back many years and is closely wrapped up with the duty which is levied on motor fuel, (which is, incidentally also charged on 'spirits used for making power methylated spirits'). But at present it may be reckoned that a bona fide industrial user can obtain duty-free light hydrocarbon oils either as a bonded user, or, as a user licensed to claim back from the Customs and Excise the duty paid to supplier; in either case it is necessary to satisfy the Commissioners that the use is 'approved' in law and that the records kept satisfy the requirements laid down.

Samples and small quantities may frequently be sent by post and it is again important to understand the legal position. Any liquid flashing below 90° F (32.2° C) is completely banned − hence products falling within the meaning of either the Petroleum Act or The Cellulose Regulations are forbidden, and products based on, e.g., ethyl or isopropyl alcohol, are also illegal. Products flashing between 90° F (32.2° C) and 150° F (65.5° C) may be posted in containers no larger than 2 pints in size and specially packed. (This usually means one container inside another with absorbant packing material around the first to mop up in the event of spill with 7½% ullage.) For transport by British Railways it is difficult to express the regulations in simple terms since solvents or products containing solvents might be classified as 'inflammable liquids', 'dangerous chemicals', or just 'miscellaneous'. Inflammable liquids are grouped as Class A (flash p. below 73° F (22.8° C) and liquids completely miscible with water and compositions made therewith with flash p. below 73° F (22.8° C)), Class B, (flash p. between 73° F − 150° F (22.8° C − 65.5° C). It is becoming current practice to label goods with the pictorial labels described earlier so that some degree of coherence in definition may be traced between rail and water transport requirements. A further complication arises in so far as the carriage of 'Dangerous Goods' is either rigidly controlled or completely forbidden over certain sections of railway as well as in railway shipping and ferries. The complexity of such Regulations renders it essential that they be consulted for each particular case; generalizations in interpretation can be dangerous.

References

[1] Draft Statutory Instruments, No. 196 *Factories – The Highly Flammable Liquids.* Regulation 196.
[2] S.R.&O. 1902, No. 623
[3] S.R.&O. 1922, No. 329
[4] Petroleum (Consolidation) Act 1928, (18&19 Geo 5. C. 32)
[5] Fire Protection Association. Booklet No. 32 Appendix C.
[6] S.R.&O. 1926, No. 1422
[7] S.R.&O. 1947, No. 1443
[8] S.I. 1948, No. 2373
[9] S.I. 1948, No. 1758
[10] cf. Institute of Petroleum 33/35
[11] A S T M D 50–49
[12] A S T M D 93–49
[13] S.R.&O. 1934, No. 990
[14] *The Carriage of Dangerous Goods and Explosives in Ships,* H.M.S.O. 1961
[15] The Finance Act 1964, S. 6 (1) (2).
[16] Notice by the Commissioners of Customs and Excise. No. 179 1967.

CHAPTER EIGHT

Toxicity

It is prudent to assume that the use of any organic solvent involves some hazard to health. This may result from accidental aspiration of the liquid, from contact of the liquid with the skin or eyes, or from inhalation of its vapours. In fact, it may be assumed that the vapours of all volatile substances are toxic if they are inhaled in a sufficiently concentrated state for a sufficient length of time. Two types of effects should be anticipated; acute reactions which can result from a short exposure to relatively high concentrations and chronic effects following repeated and prolonged exposure to low concentrations. Acute poisoning is usually more dramatic in its effects. The consequences of chronic poisoning develop more gradually and, if not recognized in time, may lead to permanent organ damage. However, it must be emphasized that it is frequently difficult to distinguish between acute and chronic effects.

The probability of a lethal atmospheric concentration being reached depends on the volatility of the substance. However, local conditions of temperature and relative humidity may result in an increased rate of evaporation.

Although toxic quantities of some solvents may be absorbed through the skin, local and severe damage may occur as a result of contact of the liquid or of its vapour with the skin. This may involve removal of the superficial layers of fat, thus exposing the skin to infection by micro-organisms. This occurs very commonly with those solvents which dissolve animal fats, particularly aromatic hydrocarbons, chlorinated aliphatic hydrocarbons, esters, ethers and ketones but not alcohols and glycols. More severe damage, including blistering and dermatitis, may also occur although individuals vary considerably in their sensitivity in respect of the latter. The action of all types of solvents on the skin can be reduced by the use of suitable barrier creams and by close attention to personal hygiene.

It has been thought necessary to provide some guidance concerning the nature of the hazards involved with some industrial organic solvents. Because of their low volatility, plasticizers and high-boiling solvents have been excluded. While much of our knowledge concerning the toxic hazards of solvents has been obtained from studies with experimental animals, studies in which humans have been exposed to carefully controlled atmospheres are by no means rare and, of course, much information has been obtained from the examination of personnel employed in industry. The interpretation of the results of animal experimentation should be treated with caution because the response of different species can vary quite considerably. Attempts have been made to establish limits for the airborne concentrations of many chemicals under which it is believed that nearly all workers may be repeatedly exposed day after day without adverse reaction. Because of wide variation in individual susceptibility, a small percentage of workers may, however, experience discomfort from some substances at concentrations on or below the threshold limit.

These values, which are expressed as parts of vapour or gas per million parts (ppm) of contaminated air by volume at 25° C and 760 mm Hg pressure, are referred to as Threshold Limit Values (TLV), although the term Maximum Allowable Concentration (MAC) is frequently used in the earlier literature. The values currently adopted in Great Britain are those established by the American Conference of Governmental Industrial Hygienists (ACGIH). The values should be used as guide lines only and should not be used as fine lines between safe and dangerous concentrations. The complete list has been reproduced by the Department of Employment and Productivity under the title Dust and Fumes in Factory Atmospheres (1968), New Series No 8, and is available from Her Majesty's Stationery Office.

Aromatic hydrocarbons

The liquid aromatic hydrocarbons are irritants which, on repeated or prolonged contact with the skin, may cause dermatitis. Severe lung damage can result from aspiration of small amounts of these substances. The vapours are irritating to mucous membranes and systemic injury can result from their inhalation.

Benzene

Industrial benzene poisoning occurs almost exclusively from inhalation of the vapour, although small quantities of liquid benzene can be absorbed through the skin. Acute benzene poisoning is due to its narcotic action; a concentration of 7500 ppm is dangerous for exposures exceeding 30 minutes, while 20,000 ppm is lethal. Systemic injury may occur as a result of brief exposures to high atmospheric concentrations or prolonged inhalation of low concentrations of benzene vapour. Chronic benzene poisoning is characterized by its effects on the blood-forming organs, leading to severe anaemia. Wherever possible, benzene should be replaced as a solvent by one that is free from this particular hazard. Chronic benzene poisoning is a notifiable industrial disease under Section 82 of the Factories Act, 1961, Order No 1505, December 1924.

Toluene

Although at high concentrations toluene is considered to be more dangerous than benzene, the hazards arising from repeated exposures to low concentrations are substantially less. Daily exposures to 200 ppm toluene produce fatigue, weakness and mental confusion, while nausea, dizziness and headache occur at 600 ppm. No effects have been observed following exposure to 100 ppm. The effects of toluene on the skin are considered to be more severe than those of benzene and dermatitis may occur as a result of prolonged skin contact. Toluene, like the other alkylbenzenes, does not affect the blood-forming tissues.

Ethyl benzene

No organ damage has been reported following prolonged and repeated exposure to 2200 ppm ethyl benzene. A concentration of 5000 ppm causes severe irritation and can be only briefly tolerated. No systemic injury has been reported from its industrial use.

Xylene

Although the various isomers of xylene differ in their acute toxicity, it appears that they are all more acutely toxic than benzene or toluene. Repeated exposure to non-narcotic concentrations causes nausea and vomiting. In most respects, the effects produced by xylene are similar to those of toluene. Skin irritation from xylene is more serious than from either benzene or toluene. A concentration of 200 ppm xylene causes appreciable irritation.

Styrene

A concentration of 10,000 ppm is dangerous for exposures exceeding 30 minutes. Immediate and severe eye irritation followed exposure to 2000 ppm. Styrene vapour in concentrations of 200 to 400 ppm has a transient irritating effect on the eyes and nose. As with the other aromatic hydrocarbons, dermatitis may occur from prolonged skin contact. Repeated exposure to styrene vapour may cause headache, insomnia and induce a state akin to drunkenness.

Cumene
Liquid cumene is absorbed through the skin more rapidly than toluene, xylene or ethyl benzene. Repeated inhalation of 2000 ppm of the vapour produces mild narcosis. Cumene has no deleterious action on the blood.

p-tert-Butyl toluene
Exposure to 934 ppm p-tert-butyl toluene for 60 minutes can be fatal. Repeated and prolonged exposure to 25 ppm is without effect, while moderate liver and lung damage may occur at 50 ppm. p-tert-butyl toluene is more potent than the other aromatic hydrocarbons in its effects on the central nervous system.

Tetrahydronaphthalene
Repeated exposures to 275 ppm tetralin cause severe damage to kidneys, liver and lungs. Like the other aromatic hydrocarbons tetralin is an irritant to skin and eyes and possesses dermatitic activity.

Chlorinated aliphatic hydrocarbons
The toxic properties of the individual substances vary considerably, particularly with regard to their effects on the liver and kidney, and in their narcotic potency.

Methylene dichloride (dichloromethane)
This substance is considered the least toxic of the four chlorinated methanes. Deep narcosis can be produced within 30 minutes by exposure to 20,000 ppm. Headache, giddiness and irritability occur at 2000 ppm. Methylene dichloride is only mildly irritating to the skin on repeated contact.

Chloroform (trichloromethane)
The most outstanding effect from acute exposure to high concentrations of chloroform is depression of the central nervous system. Repeated exposure to concentrations of 77 ppm chloroform causes vomiting, while dizziness occurs after exposure to 1000 ppm for 7 minutes. Chronic exposure may cause liver injury. Habituation to chloroform is not uncommon.

Carbon tetrachloride (tetrachloromethane)
Exposure to this material constitutes an important health hazard. While fatal narcotic poisoning is rare, severe symptoms of poisoning may follow a single exposure to concentrations of 1000 ppm. The effects of carbon tetrachloride are similar to those of chloroform, although its action on the liver is more pronounced. Loss of weight, jaundice, cirrhosis of the liver and kidney damage, including nephrosis and anuria, may occur as a result of continuous exposure to 50 ppm.

Ethylene dichloride (1,2-dichloroethane)
Exposure to 3000 ppm for 60 minutes is fatal. The symptoms of poisoning are largely related to central nervous system depression and gastro-intestinal upset. Liver, kidney and adrenal injuries may occur at sub-lethal concentrations, although the effects are less marked than those from comparable exposures to carbon tetrachloride. Ethylene dichloride can be absorbed through the skin; prolonged skin contact causes severe irritation.

Methyl chloroform (1,1,1-trichloroethane)
This is one of the least toxic of the group of chlorinated aliphatic hydrocarbons. Humans exposed to 900 ppm of the vapour experienced transient mild irritation. Above 1700 ppm signs of "drunkenness" have been observed. No physiological effects have been noted at concentrations below 350 ppm. The fatal atmospheric concentration for humans is in excess of 50,000 ppm. It is unlikely that significant organ injury will result from repeated exposures in the absence of any acute effects. Repeated contact of the liquid with the skin results in slight irritation secondary to the solvent's defatting action.

Acetylene tetrachloride (1,1,2,2-tetrachloroethane)
This substance is considered to be the most dangerously toxic of the chlorinated aliphatic hydrocarbons. It is a powerful narcotic and causes profound metabolic injury. The symptoms of poisoning include jaundice and toxaemia. The main organs affected are the liver, the kidneys and the lungs. It is a skin irritant and may cause severe dermatitis. The fatal concentration for short exposure is estimated at greater than 3000 ppm. Toxic jaundice caused by contact with tetrachloroethane is notifiable under the Factories Act, 1961, Section 82, Order No 1170 of November 1915.

Pentachloroethane
A narcotic poison of considerable potency, it is severely irritating to the eyes and upper respiratory tract. It has a marked effect on the liver, similar to that of tetrachloroethane. It is used only infrequently and no effects in humans have been recorded.

Trichloroethylene
This substance is a narcotic of considerable potency and death may occur as a result of respiratory failure if exposure is severe and prolonged. Although there is little evidence that trichloroethylene has a marked chronic action on either the liver or the kidneys, deaths have occurred following prolonged exposure, probably caused by ventricular arrythmia. Dermatitis may result from repeated contact with the liquid. Voluntary habituation has been reported.

Tetrachloroethylene
Although chronic exposure may result in injury to the liver and kidneys, the primary hazard is that due to its narcotic action. It is considered to be among the less toxic of the industrial aliphatic chlorinated hydrocarbons. Subjective symptoms have been recorded in men exposed repeatedly and continuously to 100 ppm tetrachloroethylene.

Ketones

On the whole, the hazard from the industrial use of ketones is not great. They are to some extent narcotic. As high concentrations cause severe irritation this effect acts as a pre-narcotic warning. With repeated exposures, the most common symptoms are headache, drowsiness and nausea.

Acetone
This is one of the least dangerous of industrial organic solvents. The inhalation hazard is negligible so narcosis only occurs at high concentrations. Repeated exposures to 2000 ppm is without effect.

Methyl ethyl ketone
Slight irritation of the throat occurs at 100 ppm and of the eyes at 200 ppm. Concentrations above 300 ppm cause headache and nausea. No serious organ damage and no severe narcotic effects have been recorded from the industrial use of methyl ethyl ketone.

Mesityl oxide
The maximum vapour concentration that can be tolerated for 60 minutes without serious effect is 1000 ppm. It causes injury to kidneys, liver and lungs. If irritative warning properties are not ignored, no hazards from inhalation should occur. The lowest concentration causing irritation is 25 ppm. Sustained or repeated contact of the liquid with the skin may cause dermatitis.

Isophorone
Irritation of eyes, nose and throat occurs at 10 ppm of the vapour. Corneal opacity has been reported after a 4-hour exposure to 840 ppm. Death is the result either of narcosis or lung irritation.

Cyclohexanone
Irritation of eyes, nose and throat occurs at 75 ppm. Slight kidney and liver injury may occur after repeated exposures to 300 ppm. Repeated skin contact may cause dermatitis.

Glycol derivatives
Ethylene glycol monomethyl ether (methyl cellosolve)
Irritation of the respiratory tract and lungs and severe kidney damage occurs following repeated exposures to 800 ppm of methyl cellosolve. The material is readily absorbed through the skin and systemic poisoning may occur. The material principally acts on the kidney, brain and blood and there is little doubt that the material should be considered highly toxic.

Ethylene glycol monoethyl ether (ethyl cellosolve)
Experiments with animals indicate that this material is distinctly less toxic than the corresponding methyl ether; the absence of any effect under industrial usage substantiates this conclusion.

Ethylene glycol monobutyl ether (butyl cellosolve)
Respiratory and eye irritation, narcosis and damage to kidney and liver will probably occur after exposures to a concentration of 300 ppm and above. The first sign of abnormality in man resulting from excessive exposure would be an abnormal blood picture together with the excretion of haemoglobin in the urine. More intense exposure would be likely to cause fragility of erythrocytes and haematuria. Ethylene glycol monobutyl ether is unusual because of its haemolytic action. Because

of its low vapour pressure the hazard from skin absorption should not be ignored.

Aliphatic esters

At high concentrations, the vapours of all simple aliphatic esters will produce narcosis. In general, the anaesthetic potency increases in the order methyl < ethyl < propyl < butyl. The onset of, and recovery from, anaesthesia is much slower than with the chlorinated aliphatic hydrocarbons. Most aliphatic esters, the formates particularly, are irritant. This may be a property of the ester or of the acid liberated by hydrolysis of the ester in the organism. Cumulative effects are not marked either in man or in animals.

Methyl formate

This material is considered to be the most hazardous of the aliphatic esters. Although a weaker narcotic than some of the other esters, prolonged inhalation of high concentrations may result in convulsions and coma. Methyl formate affects the lungs, causing congestion and oedema. The irritant properties of methyl formate do not provide adequate warning of possible harmful conditions. Exposure to 3500 ppm for 30 minutes or 1500 ppm for 180 minutes is without effect.

Ethyl formate

This causes less irritation than the methyl ester and is less likely to cause respiratory disturbance. A concentration of 330 ppm is irritating to eyes and throat. It is a more powerful narcotic but does not possess the convulsant activity of methyl formate.

Butyl formate

Concentrations of 10,300 ppm cause immediate severe irritation of the eyes, becoming intolerable within 1 minute.

Amyl formate

The narcotic potency of amyl formate is about three times that of amyl acetate.

Methyl acetate

Marked irritation of the eyes, nose and throat is experienced following exposure to a concentration of 8000 ppm. Part of the toxic action of methyl acetate, particularly in respect of effects on the optic nerve, may be due to methyl alcohol liberated by hydrolysis of the ester.

Ethyl acetate
The irritant properties of ethyl acetate are less than those of propyl, amyl and butyl acetate. It has been stated to cause hypersensitivity in skin and mucous membranes, and that this may result in a tendency to eczema. The narcotic concentration is greater than 5000 ppm.

Butyl acetate
Slight irritation may be experienced at 200 ppm. The narcotic concentration is about 10,000 ppm.

Ethers

Ethyl ether
The primary effect of ethyl ether is on the central nervous system. It is a safer anaesthetic than chloroform. Repeated exposures to concentrations below the flammability level cause headache, dizziness and loss of appetite. Ethyl ether is irritant to the eyes but not to the skin.

Isopropyl ether
Primarily an anaesthetic, isopropyl ether is about twice as toxic as ethyl ether. Relatively brief exposure to 800 ppm causes irritation of eyes and nose and a degree of respiratory discomfort.

Dichloroethyl ether
Although dichloroethyl ether in high concentrations is a narcotic, the severe irritation resulting from non-narcotic concentrations above 500 ppm can be tolerated for only a few minutes. At this concentration coughing and nausea may also be experienced. Repeated exposure to 35 ppm dichloroethyl ether is without effect. No cases of injury from its industrial use have been reported.

Aliphatic alcohols

In general, the inhalation hazards presented by aliphatic alcohols is low. They are weaker narcotics than the corresponding hydrocarbons and chronic effects are unlikely to occur, with the possible exception of methyl alcohol.

Methyl alcohol
Inhalation of concentrations greater than 40,000 ppm may be fatal due to depression of the central nervous system. Lower concentrations may cause headache, dizziness, fatigue and gastro-intestinal disturbance.

Toxic quantities may also be absorbed through the skin. Exposure to 1000 ppm for 30 minutes is unlikely to be harmful. Permanent eye injury can be caused by the inhalation of methyl alcohol.

Ethyl alcohol

Although more acutely toxic than methyl alcohol, its effects are more transient as it is oxidised in the body to carbon dioxide and water. Slight headache occurs after 30 minutes' exposure to 1400 ppm ethyl alcohol. The effects of prolonged exposure to high concentrations include irritation, headache, fatigue, nausea and narcosis.

n-Butyl alcohol

The use of this substance at ambient temperature is unlikely to present any toxic hazard. Moderate eye irritation occurs at 200 ppm. No systemic injury was reported after prolonged and repeated exposures to 100 ppm. Dermatitis may occur following prolonged contact of the liquid with the skin.

Ethylene chlorohydrin (2-chloroethanol)

Absorption by any route, including the skin, may lead to severe injury or death. The vapour causes irritation of mucous membranes, nausea, vomiting, numbness, respiratory failure and visual disturbance at high concentrations. Death results from injury to the lungs or to the brain. Exposure to concentrations above 305 ppm may be fatal.

Miscellaneous

Carbon disulphide

A potent narcotic, it is also a severe nerve poison. Toxic effects have been observed following repeated exposure to 37 ppm carbon disulphide. Exposure to 4800 ppm is fatal. Serious injury has been observed in men exposed to 1150 ppm for 30 minutes. The material has an irritant effect on the lungs. Dermatitis may result from contact of the vapour or liquid with the skin. Poisoning from carbon disulphide is notifiable under the Factories Act, 1961, Section 82, Order No 1505 of December 1924.

Nitropropane

1- and 2-nitropropane are irritants and severe liver poisons. In this respect they are about as toxic as carbon tetrachloride. They are more toxic than either nitromethane or nitroethane.

Furfural

A relatively strong skin irritant and may cause dermatitis. Death from pulmonary oedema has been observed following inhalation of 2800 ppm for 30 minutes. The inhalation hazard is low because of its relatively low volatility. A concentration of 2 ppm can produce irritation and headache.

Dioxane

The vapour of this material has poor warning properties and can be inhaled in amounts that can cause severe damage to liver and kidneys. Brief exposure to 300 ppm dioxan produces mild irritation of eyes, nose and throat. Severe injury will occur following repeated and prolonged exposures to 1000 ppm. The liquid can be absorbed through the skin in amounts sufficient to produce organ damage.

Tetrahydrofuran

This substance is a more potent narcotic than the various glycol derivatives. Although it causes injury to the kidneys, the damage is not so pronounced as that caused by dioxan. Concentrations above 3000 ppm cause severe irritation. Workers repeatedly exposed to the vapours of tetrahydrofuran may experience severe headaches.

Petroleum spirits

The toxicity of petroleum fractions varies considerably with composition; although only approximate, it may be accepted that the toxicity of a petroleum fraction with a volatility similar to that of benzol is about one-half that of benzol; the acute effects of petroleum are attributed mainly to the cyclo-paraffins, although the poisoning, whether acute or chronic, is essentially one of nerve poisoning. Prolonged contact may give rise to anaemia. The fatal concentration for short exposure to gasoline is estimated at about 20,000 ppm.

White spirit

Dangerous concentrations are unlikely to occur under normal conditions of use because of the low volatility of the material. The vapour can cause giddiness, but recovery is rapid and there is no evidence of any residual effects. The only significant hazard is that of skin irritation.

Turpentine

Exposure to concentrations of 720 – 1100 ppm results in mucous membrane irritation (particularly of the eyes), headache, nausea and chest pains. Transient kidney damage may also be experienced. Some individuals may develop a hypersensitivity to turpentine after prolonged and repeated exposure to the material.

The information in this chapter has been of necessity somewhat limited. The following publications provide comprehensive and authoritative data concerning the toxic hazards of a wide variety of organic solvents and chemicals.

Browning, E., *Toxicity and Metabolism of Industrial Solvents,* 3rd edition. (Elsevier Publishing Company, Amsterdam – London – New York) 1965.
Patty, F. A., (Ed), *Industrial Hygiene and Toxicology,* 2nd revised edition. *Vol II. Toxicology.* (Interscience Publishers, New York – London) 1962.

PART II

PART II

Introduction

The figures quoted in this section for the physical and chemical characteristics of a solvent, if not stated otherwise, refer primarily to those obtaining for a technical product of good quality. It should be borne in mind that the products emanating from different sources may vary in their purity, and the figures given should in general be regarded as the permissible limits rather than those usually occurring although many products are now in accordance with agreed and recognised standards. The figures relating to the characteristics of the pure substance frequently follow those of the technical product in order that some idea can be obtained regarding the purity of the latter. It should be borne in mind that the quality or composition of a solvent may be varied to suit trade conditions, but the trend is generally towards greater purity.

In many cases it is not desirable or expedient that a chemically pure substance should be used for lacquers. In some instances the presence of 'impurities' considerably enhances the desirable properties of a solvent. In other cases the cost of removing certain impurities may be out of proportion to the advantages to be gained.

Specifications of an official nature which are quoted are often incomplete and are not given verbatim, but in a convenient abbreviated form suitable for quick reference only; if more definite details are desired, the original specifications should be consulted [1].

An effort has been made to make this section as complete as possible, and every source of information available has been freely made use of; it has not always been possible to acknowledge all of these. Both the author and the reviser have also drawn largely from their own experience. Particular acknowledgement should be made to the following firms and institutions for their assistance:

Albright and Wilson Ltd.
American Society for Testing Materials
B.P. Chemicals Ltd.
British Standards Institution
Carless Capel & Leonard Ltd.
Commercial Solvents (GB) Ltd.
Geigy Co. Ltd.
General Metallurgical and Chemicals Ltd.
Glyco Products Co. Inc.
Hercules Powder Co. Inc.

Imperial Chemical Industries Ltd.

Monsanto Chemicals Ltd.

Quaker Oats Co.

Bush Boake Allen - division of Albright & Wilson Ltd.

Sharples Chemicals Inc.

Shell International Chemical Co. Ltd.

US Industrial Chemicals Inc.

In view of the large and increasing number of proprietary names they have mostly been eliminated from the text and gathered in Appendix I. These names are in many instances trade marks and the composition of the solvents or other chemicals to which these names refer is that generally quoted in various journals or catalogues. These compositions are given here in good faith but their accuracy cannot be guaranteed [2].

References

[1] British Standard Specifications are abstracted by permission of the British Standards Institution, 2 Park Street, London W 1, from whom official copies of the full specifications can be obtained. National Benzole Association Specifications can be obtained in full detail from The National Benzole Association, Wellington House, Buckingham Gate, London S W 1.
ASTM Specifications are abstracted by permission of the American Society for Testing Materials, 1916 Race Street, Philadelphia 3, USA:
All the specifications are revised from time to time and accordingly may need verification.

[2] See: Van Hock, *Farber Zeitung*, 1927, **32**, 1737; Noll, *ibid*, 1927, **32**, 1553; Noll, *Farbe und Lac.*, 1926, **31**, 6, 92, 172; Main W., *Enduits Cellulosiques*, 1930, Paris; Jordan O., *Losungsmittel*, 1932, Berlin; Bianchi-Weihe, *Cellulose-ester Lacke*, 1931, Berlin; Ulrich H., *Decknamen und chem. Zusammensetzung*, 1935, Hamburg; Munziger W., *Technologie d. Weichmachungsmittel*, 1935, Munich; Durrans-Merz *Losungsmittel und Weichmachungs Mittel*, 1933, Halle; Gardner, *Physical and Chemical Examination of Paints, etc.*, Washington; US Public Health Service *Public Health Reports*, 1946, **61**(1), 132.

Abbreviations and notes

sp. gr.	Specific gravity at 15.5° C (60° F), compared with water at the same temperature unless otherwise stated.
b. r.	Boiling range.
b. p.	Boiling point.
v. p.	Vapour pressure in mm of mercury at 20° C.
flash p.	Flash point.
ign. p.	Permanent ignition point.
s. p.	Solidification point
m. p.	Melting point.
evap. period	Evaporation period (ether = 1)
n_D	Refractive index for the D line of sodium (at 20° C unless otherwise stated)
lat. ht.	Latent heat of vaporization in gram calories/gram at b. p.
sp. ht.	Specific heat in gram calories/gram at about 20° C
ht. comb.	Heat of combustion in gram calories/gram.
K-B value	Kauri-butanol value.
aniline p.	Aniline point
dil. ratio	Dilution ratio for nitrocellulose unless otherwise stated.
cub. expn.	Coefficient of cubical expansion/° C at about 20° C.
s. ten.	Surface tension in dynes/cm at 20° C
visc.	Dynamic viscosity in centipoises at 20° C†
therm. cond.	Thermal conductivity in gram calories/cm² /s/deg C.
elec. cond.	Electric conductivity in reciprocal ohms at 25° C*
dielec. const.	Dielectric constant at ∞ frequency at 20° C
crit. temp.	Critical temperature
expl. lim.	Explosive limits in percentages by volume in air

† To convert kinematic viscosity in centistokes divide by the density of the liquid.

* Greatly affected by impurities

CHAPTER ONE

Hydrocarbons and sundry solvents

Benzene

Benzene (or benzol) is a member of the closed ring group of aromatic solvents commercially derived from both coal tar and petroleum sources; it must not be confused with benzine which mainly contains the petroleum straight-chain paraffins hexane and heptane. Economics can exert a considerable effect upon the source of benzene supply in a particular locality and it is important to know the source, since it can affect impurities likely to be present. Petroleum-derived material is invariably of very high purity, partly on account of its high degree of refinement (which follows closely that accorded to modern motor fuels) and partly because its purity is an essential part of its later use as a petrochemical intermediate, eg for production of ethyl benzene, acetophenone, phenyl ethyl alcohol and styrene. Other commercially important derivatives of benzene are phenol, detergent alkylates, and various insecticides. The manufacture of these latter creates periodic shortages of benzene and some is therefore synthesized from toluene to help fill the gap; it is unlikely that any great quantity of synthetic benzene finds its way into the solvent market. Depending on its source the trace impurities which may occur in benzene are toluol, xylols, carbon disulphide, thiophene, acetonitrile, and pyridine.

Benzene can be used as a diluent in lacquers or adhesives based on cellulose nitrate, cellulose aceto-butyrate, cellulose acetate, vinyl copolymers and some acrylics; it may be used as a prime solvent in ethyl cellulose, styrene, and styrene copolymer, some acrylics and rubber-based products. It is a valuable constituent of thinners intended for use with any of the foregoing and it is a solvent for most oils, fats, waxes, alkyd resins, and many natural resins; it is not a solvent for shellac. It is a solvent for a number of other cellulose esters and ethers, in particular, cellulose dinaphthenate, dilaurate, dipalmitate, distearate, dinitrolaurate, dinitropalmitate, diacetyl laurate, and diacetyl palmitate [1]. If mixed with alcohol it will dissolve benzyl cellulose and other ethers. It is a solvent for most oils and for ester gum, benzyl abietate, copal ester, polystyrene, polyisobutylene, polyvinyl acetate, colophony, elemi, fluid silicones, vegetable oils, phosphorus, iodine and gamma-benzene hexachloride 29%, coumarone, benzyl resin, mastic, raw rubber, gutta percha resin, thio-urea resins, chlornaphthalene resin, and naphthalene formaldehyde resin. It does not dissolve cellulose acetate or nitrate, sandarac, manila or copals, and is non-miscible with water, aqueous solvents and glycols, monochlorhydrine, monacetin. Cellulose acetate absorbs 22% by weight of benzene at 25° C. It attacks polythene, polyvinyl chloride, and methyl methacrylate but not nylon. Benzene is available in several grades.

N.B. The Benzene Solvents (Limitations of Use) Regulations which are being made under Section 76 of the Factories Act 1961, will prohibit in factories and other places to which the Factories Act 1961 applies the use as a solvent of benzene or any liquid containing more than 1% by volume of benzene. The prohibition will not apply where such use is necessary in effecting a chemical synthesis or in the case of any process carried out within a totally enclosed system or in the case of any laboratory process in which such use is unavoidable.

British Standard Specification, BS 135/1:1963, for pure benzene, requires: sp. gr. 0.880–0.886; b.r. 1% to 96% over range of 1 deg C, this range to include 80.1° C; max residue 5 mg per 100 ml; neutral; total sulphur 0.2% max.; sulphur as CS_2 0.1% max.; free from mercaptans, H_2S and undissolved water; colour, acid-wash and free sulphur tests.

(National Benzole and Allied Products Association 1:1960)

British Standard Specification, BS 135/3:1963 for 90's benzene, requires: sp. gr. 8.874–0.884; at least 90% but not more than 95% collected at 100° C; max. residue 5 mg per 100 ml; neutral; total sulphur 0.4% max.; free from mercaptans, H_2S and undissolved water; colour, acid-wash and free sulphur tests.

(National Benzole and Allied Products Association 4:1960)

ASTM Specification, D836-50 for industrial grade benzene requires: sp. gr. 0.875–0.886; total distillation range not more than 2 deg C including 80.1; acidity–nil; free of H_2S and SO_2; colour, acid wash, and copper corrosion tests.

Chemically-pure Benzene.

sp. gr. 0.8789_{20}; m. p. $5.5°$ C; b. p. $80.07°$ C; v. p. 75; flash p. $12°$ F $(-11°$ C). r. i. 1.501; sp. ht. 0.408; lat. ht. 94; elec. cond. 5×10^{-17}; dielec. const. 2.3 at $15°$ C; K-B value 100; cub. exp. 0.00124; vis. 0.69_{15}, 0.66, 0.57_{30}; s. ten. 32. ht. comb. 9960; Water dissolves 0.08%; dissolves 0.06% of water.

The following azeotropic mixtures are known:

Benzene %			b. p. ($°$ C)
55	Cyclohexane	45%	77.5
5	n-Hexane	95%	68.5
12	Me-Cyclopentane	88%	71.4
99	n-Heptane	1%	80.1
60	Methyl alcohol	40%	58.3
68	Ethyl alcohol	32%	68.2
83	n-Propyl alcohol	17%	77.1
67	iso-Propyl alcohol	33%	71.9
91	iso-Butyl alcohol	9%	79.8
84	sec-Butyl alcohol	16%	78.8
63	tert-Butyl alcohol	37%	74.0
6	Ethyl Acetate	94%	77.0
62	Methyl ethyl ketone	38%	78.3
Benzene 39%	Cyclohexane 52%	Methyl Cellosolve 9%	b. p. 73

Toluene

Toluene or Toluol is extensively used either as a diluent or as a solvent in a vast range of coatings. It also finds application as a prime solvent in adhesives where high aromatically and rapid release from the applied film are important characteristics. It is obtained from both coal and petroleum sources, the latter material usually containing some quantity of methyl cyclopentane b. p. $117-120°$ C. The odour is not especially pleasant (reminiscent of onions) and the material is narcotic; many workers regard it as less dangerous than benzene.

Toluene is a solvent for ethyl cellulose, ester gums, resin and its many variations, epoxy resins, amino resins, alkyd resins, cyclo-hexanane resins, sulphinamide resins, polyester resins, gum copal ester, raw rubber; chlorinated, cyclized and semerized rubber, butadiene/styrene acrylonitrile rubber, gutta percha resin and mastic, and most acrylic or alkyd methacrylic resins, polystyrene, polyvinyl acetate, polyisobutylene, chlorinated polyvinyl chloride, chlorinated poly-propylene, chlorinated diphenyl and chlorinated paraffin, natural synthetic waxes, oils, bitumen gilsenate. It is not a solvent for shellac, sandarac, polyvinyl chloride, polyethylene cellulose acetate and cellulose nitrate, nylon, polytetrafluorethylene and moulded phenolic resins. It is miscible with oils, hydrocarbons, and other solvents including pure ethyl alcohol, but not with industrial methylated spirits (except in limited ratios), monochlorhydrin, glycol and water; it can be emulsified with water, with a wide range of emulsifying agents, or with hydrated isopropanol (due to water present in each case).

Petroleum-derived toluene is becoming increasingly important as a chemical intermediate for the manufacture of TNT and other important aromatic derivatives (dyestuffs, pharmaceuticals) [2].

Characteristics (Pure): sp. gr. 0.872, 0.867_{20}; b. p. $110.56°$ C; flash p. $40°$ F ($4°$ C); ign. p. $21°$ C; m. p. $-95.2°$ C; v. p. 22; n_D 1.4966; lat. ht. 86; sp. ht. 0.40; ht. comb. 10150; cub. expn. 0.00107; visc. 0.696_{15}, 0.566_{30}; s. ten. 28; elec. cond. 1.4×10^{-14}; dielec. const. 2.38; water dissolves 0.05%; dissolves 0.04% water.

British Standard Specification, BS 805/1: 1963 for toluene requires: sp. gr. 0.865–0.872; b. r. (1%–96%) not to exceed 1 deg C and to include $110.6°$ C, max. total sulphur 0.2%; neutral; max. residue 5 mg per 100 ml; free from mercaptans, H_2S and water, tests for colour, acid-wash, and free sulphur.

(National Benzole and Allied Products Association 6A:1960)

British Standard Specification, BS 805/4: 1963 for 90's toluene, requires: sp. gr. 0.860–0.875; initial b. p. at least $100°$ C and at least 90% taken by $120°$ C; neutral; max. residue 5 mg per 100 ml; free from mercaptans, H_2S, and undissolved water; tests for colour, acid-wash and free sulphur.

(National Benzole and Allied Products Association 8A:1960)

ASTM Specification, D362-65 for industrial toluene requires: sp. gr. 0.864–0.874; b. r. not more than 2 deg C from initial b. p. to dry p. including $110.6°$ C; acidity not more than 0.005% by weight; free of H_2S, undissolved water and SO_2, tests for colour, acid wash, and corrosion.

Toluene obtained from petroleum is normally of nitration grade quality conforming to BS 805/3: 1963 and ASTM 841–66.

The rate of evaporation of toluene is about four times that of n-butyl alcohol and about twice that of butyl acetate.

The following azeotropic mixtures are known:

Toluene (%)			b. p. ($^{\circ}$ C)
32	Ethyl alcohol	68%	76.7
51	n-Propyl alcohol	49%	92.6
31	Isopropyl alcohol	69%	80.6
68	n-Butyl alcohol	32%	105.5
56	Isobutyl alcohol	44%	101.1
74	Epichlorhydrine	26%	108.3

and a ternary mixture consisting of toluene 50%, ethyl alcohol 38%, water 12%, boiling at 74.6° C [13].

Xylene $C_6H_4(CH_3)_2$

Commercial xylene invariably comprises a mixture of the three isomers, *ortho-*, *meta-* and *para-* and it is obtained from both coal tar and petroleum sources; it is possible to prepare specific distillation ranges and the most common are 2, 3, and 5 degree material, which simply means that 5 − 95% will distil over within 2, 3, or 5 deg C. Basic properties of the three constituent isomers are as follows

	Ortho.	Meta.	Para.
sp. gr. at 20° C	0.8801	0.8641	0.8610
b. p. ($^{\circ}$ C)	144.2	139.1	138.3
s. p. ($^{\circ}$C)	−25	−48	13
n	1.5052	1.4972	1.4958
cub. expn.	−	0.00010	0.00102
v. p.	29	31	32
s. ten.	−	32.2	28.3
sp. ht. (30° C)	−	0.387	0.397
lat. ht.	−	81.8	81.0
Visc.	0.810	0.620	0.648
dielec. const. (at 30° C)	2.57	2.35	2.26
K-B value	107	103	97
aniline p.	8.4	10.0	10.6

As far as practical solvency characteristics are concerned the percentage of each isomer present in the final blend has little effect, but trace impurities may need detection; thus the presence of toluene can be detected by flash-point testing and that of trimethyl benzene (usually 1, 2, 4,-trimethylbenzene or pseudocumene) may be pinpointed by checking the evaporation point of a particular sample.

Xylene is not a solvent for cellulose nitrate, acetate, aceto-propionate or butyrate, polyvinyl chloride, shellac, polyester resins, polyvinyl butyral or acetal, polyethylene or nylon. Some grades of cellulose nitrate can be dispersed in specific mixtures with anhydrous ethyl alcohol, and cellulose acetate of the soluble variety (lower acetyl content) absorbs about 5% of meta-xylene at 25° C whilst flexible polyvinyl chloride can also be softened by prolonged contact with xylene.

Xylene is a solvent for ethyl and benzyl cellulose, ester gum, copal ester, benzyl abietate, rubber, gutta percha resin, polystyrene, methyl methacrylate, polyisobutylene, rubber chloride dammar, elemi, coumarone, colophony, fluid silicones, melamine-formaldehyde, and urea–formaldehyde resins, castor, linseed, and other oils. It attacks polyethylene and polyvinyl chloride but not nylon; it is a poor solvent for polyvinyl acetate and acetochloride. It dissolves 25% of gamma-benzene hexachloride at 20° C. It is miscible in all proportions with petroleum hydrocarbons and with practically all of the cellulose ester solvents with the exception of diacetin, mono-chlorhydrine, glycols, and industrial alcohol.

British Standard Specification, BS 458/2:1963 for 3° solvent xylene, requires: sp. gr. 0.860–0.875; b. r. (5%–95%) not to exceed 3.0 deg C and to lie between 137.5° C and 144.5° C; neutral; free from mercaptans, H_2S, and undissolved water; flash p. not below 22.8° C; max. residue 10 mg per 100 ml; tests for free sulphur, and colour.

(National Benzole and Allied Products Association 10B:1960)

British Standards Specification, BS 458/4:1963 for 5° solvent xylene requires: sp. gr. 0.860–0.875; b. r. (5%–95%) not to exceed 5.0 deg C and to lie between 137.0° C and 145.5° C; neutral; free from mercaptans, H_2S and undissolved water; flash p. not below 22.8° C, max. residue 10 mg per 100 ml; tests for free sulphur and colour.

(National Benzole and Allied Products Association 11A:1960)

ASTM Specifications, ASTM D 364–65 for industrial grade xylene requires: sp. gr. 0.860–0.871; min. initial b. p. 123° C, max. 5% at 130° C, min. 90% at 145° C, max. dry point 155° C; acidity not more than 0.005% by weight; free of H_2S, SO_2, and undissolved water; tests for acid wash, and corrosion.

A purified xylene of high flash-point is also available: b. r. 138°–140° C; flash p. about 85° F (30° C); ign. p. about 46° C; visc. about 0.65; lat. ht. 82; sp. ht. 0.4; auto-ignition temperature 136° C; K-B value 85–90; cub. exp. 0.001.

The following azeotropic mixtures are known:

			b. p. (°C)
o-Xylene 40%	Isoamyl alcohol	60%	128.0
m-Xylene 47%	Isoamyl alcohol	53%	127.0
m-Xylene 46%	Isoamyl alcohol	54%	136.0
p-Xylene 49%	Isoamyl alcohol	51%	126.8

Ethyl Benzene

This hydrocarbon is now available industrially. It is similar to the industrial xylenes but has the advantage of being a single substance of constant boiling point. It is colourless and resembles pure xylene in odour but is probably slightly more toxic. It is produced at about 98% purity from the Udex process coupled with extraction by aqueous diethylene glycol C_8 products of catalytic reforming. It has found some use as a solvent in certain types of adhesive for polystyrene and its compounds, and is also the starting point for the manufacture of styrene monomer (by dehydrogenation).

Characteristics
b. p. 136.2° C; sp. gr. 0.870 (0.8671 at 20° C); flash p. 68° F (20° C); m. p. −94.98° C; v. p. 15.3; visc. 0.63; n_D 1.496; dielec. const. 2.38_{30}; K-B value 98.6; aniline p. 10.0°C; sp. ht. 0.41; lat. ht. 101_{25}, 97_{50}, 91_{100}; water dissolves 0.014% at 15° C.

Solvent naphthas

Napththas have been used as solvents for many years and are obtained from both coal tar and petroleum sources. Their exact composition can vary considerably but will consist of mixtures of aromatic hydrocarbons, naphthenes and aliphatic hydrocarbons in differing proportions; naphthas may be obtained as the result of a particular distillation method or they may be obtained by blending in order to obtain, for example, specific aromatic contents. In general, they are solvents for ester gum, coumarone, bitumen, pitch, many resins and oils, plasticisers, natural and synthetic waxes, polystyrene, and rubber and its derivatives. They will not dissolve cellulose nitrate, acetate, acetate-butyrate or propionate, but, depending on their aromatic content will frequently dissolve ethyl cellulose; some blends have been prepared to similate xylene (sometimes known as Light Solvent Naphtha) and have been included in thinners for cellulose lacquers, but it is important to check the 'tail' of such naphthas since this can cause trouble (odour, cotton blush, etc.) if it is too great in quantity or too high in evaporation rate. Naphthas also find use in insecticidal sprays where their solvency for DDT and other synthetic insecticides is of importance [3].

British Standard Specification, BS 479/1:1963 for Coal-tar Solvent-Naphtha (96/160) requires: sp. gr. not less than 0.854; distillation, not more than 5% at 125° C, at least 96% at 160° C; free from H_2S, mercaptans, and undissolved water; neutral; max. residue 10 mg per 100 ml; tests for colour and acid wash.

(National Benzole and Allied Products Association 13A:1960)

British Standards Specification, BS 479/3:1963 for Coal-tar Solvent Naphtha (90/160), requires: sp. gr. not less than 0.850; distillation, not more than 5% at 125° C, at least 90% at 160° C, free from H_2S, mercaptans, and undissolved water; neutral; max. residue 10 mg per 100 ml; tests for colour and acid wash.

(National Benzole and Allied Products Association 14A;1960)

ATSM Specification D 838-50 for Refined Solvent Naphtha, requires: sp. gr. 0.850 to 0.870; distillation, not more than 5% at 130° C, at least 90% at 145° C, dry point not above 155° C; free from H_2^2S and SO_2; acidity, nil; tests for colour and acid wash.

ASTM Specification, D839-50 for Crude Light Solvent Naphtha, requires: sp. gr. 0.860 to 0.885; distillation, not more than 5% at 130° C, at least 90% at 160° C, dry point not above 180° C; acidity, nil; colour, light amber.

ASTM Specification, D840-50 for Crude Heavy Solvent Naphtha, requires: sp. gr. 0.885–0.970; distillation, not more than 5% at 150° C, at least 5% at 165° C, at least 90% at 200° C, dry point not above 220° C; acidity, nil; colour, between light amber and dark red.

Cymene

Cymene, Cymol, or p-methylisopropyl benzene has been suggested as a high boiling diluent for lacquers. It is a pleasant-smelling liquid, having an odour like that of parsley. It occurs in light resin oil and in spruce turpentine. It is obtained as a by-product in the manufacture of sulphite paper pulp, probably resulting from the dehydrogenation of pinene by sulphur, also by deliberate catalytic dehydrogenation [15].

Its physical characteristics vary with its source. Sulphite pulp cymene has sp. gr. 0.857 (at 20° C); n_D 1.49; b. p. 174–177° C. Cymene from another source [15] had sp. gr. 0.860; n_D 1.488; m. p. below −40° C; flash p. 140° F (60° C); K-B value 7.17; aniline p. −18.6° C; visc. 1 at 25° C; b. r. about 176–179° C; dielec. const. 2.25; elec. cond. 2×10^{-8}. Pure cymene has sp. gr. 0.857 at 20° C; b. p. 176.9° C; s. p. −69.8° C; n_D 1.490. It forms a constant-boiling mixture with 71% of cyclohexanol boiling at 159°C.

It is an excellent solvent for many resins and polyisobutylene but not for cellulose esters. It is miscible with oils and hydrocarbons and most of the usual solvents, but not with water. It is a stable substance and does not readily oxidize in the air, and finds use as a substitute for turpentine.

Dipentene

Dipentene is a terpene which occurs widely in volatile essential oils. It is a colourless liquid having a faint lemon odour, the pure substance has sp. gr. 0.844 at 20° C; n_D 1.472; b. p. 175–6° C. A technical product is available [4] having sp. gr. 0.845–0.860; flash p. 109° F (43° C); n_D 1.472–1.477; b. r. 175–195° C; visc. 1.54 at 25° C; K-B value 105; aniline p. −15° to 7° C. Products containing high percentages of dipentene together with similar terpene hydrocarbons are also available. Dipentene is somewhat similar to turpentine, but it oxidizes at only one-seventh of the rate of the latter. It is a solvent for rubber, rubber chloride, some bakelites and glyptals, ester gum, coumarone, colophony, waxes, metallic driers and partly for kauri. It does not dissolve cellulose acetate or nitrate. It is compatible with most of the usual resin-oil combinations for which it is a good dispersing medium, and with which it reduces 'skinning' and gelling, and also retards initial hardening.

British Standard Specification, BS 2712:1956 for Dipentene, requires: sp. gr. 0.847–0.878, 0.844–0.875$_{20\ 20}$, 0.841–0.872$_{25\ 25}$; distillation, not more than 5% below 173° C, at least 95% below 190° C; n_D 1.471–1.482; max. residue 2% by weight; neutral; flash p. not lower than 90° F (32° C); free from visible water; clear, colourless or nearly so; tests for matter in sol. in H_2SO_4 and for corrosive sulphur compounds.

Turpentine

Turpentine varies quite widely in composition, according to its source and method of manufacture. There are four types of turpentine recognised industrially, the best being gum spirit turpentine, consisting mainly of α-pinene, obtained by steam distilling the oleo-resin exudation from various coniferous trees, such as *Pinus palustris* (Long leaf pine), *Pinus caribaea* (Slash pine), and *P. heterophylla* in America, and *P. maritima* in France.

A second variety of turpentine known as wood or stump turpentine is obtained by steam-distilling the chips or stumps remaining after the pines are cut down for lumber, and a third variety from the same material by destructive distillation; these wood turpentines consist largely of dipentene. The fourth variety of industrial turpentine, known as sulphate or sulphite turpentine, is obtained as a by-product in the manufacture of paper from wood; its composition varies widely; it may consist mainly of carene (sylvestrene) or of a mixture of α-pinene with β-pinene and dipentene.

British Standard Specification, BS 244:1962, for Turpentine, Type 1, is defined as: genuine refined gum spirits of turpentine distilled from pine oleo resins; sp. gr. 0.862–0.872; distillation, not more than 1% below 150° C, at least 87% below 170° C; n_D 1.469–1.478; max. residue (all organic) not more than 2.5% by weight; flash p. not below 90° F (32° C); free from undissolved water, clear, colourless, free from undissolved matter; max. residue unpolymerisable by H_2SO_4, 11% with n_D not lower than 1.490.

British Standard Specification, BS 244:1962, for Turpentine, Type 2; requires: genuine turpentine distilled from pine oleo resins or obtained

from resinous wood by steam or destructive distillation; sp. gr.
0.859–0.875; n_D 1.463–1.483; distillation, not more than 1% below
150° C, not less than 70% below 170°C and not less than 90% below
180° C; max. residue (all organic) not more than 2% by weight; flash p.
not below 90° F (32° C); free from undissolved water, clear, colourless;
max. residue unpolymerisable by H_2SO_4, 16% with n_D not lower than
1.480.

The American Society for Testing Materials recognizes all four of the
varieties described above, the specifications being as follows (see p. 99).

Turpentine also has lat. ht. 68; sp. ht. 0.45; dielec. const. 2.26; K-B
value 50–90; visc. 1.6; auto-ignition temp. 253° C.

Rosin Spirit

Rosin spirit or rosin oil or pinolin, is obtained by the distillation of
colophony; commercial products vary considerably in composition.
Refined spirit has approximately the following characteristics: sp. gr.
0.890–0.930; b. r. 150°–300° C (25%: 150°–200° C, 20%: 200° —
250° C, 50%: 250°–270° C, 5%: 270°–300° C); n_D 1.48–1.50; flash p.
284° F (130° C).

Cyclohexane

Cyclohexane, known also as hexamethylene, naphthene, and hexa-
hydrobenzene, is a hydrocarbon somewhat similar to benzene and is
available technically [4]. It is manufactured by the catalytic hydrogenation of benzene; it occurs in Caucasian petroleum. It is a colourless,
mobile liquid, having a less pungent odour than benzene and reminiscent of carbon tetrachloride. It is probably less toxic than benzene, and
can replace benzene where the toxicity of this is objectionable.
Cyclohexane is a good solvent for shellac, soya bean oil, cottonseed oil,
groundnut oil, flaxseed oil, rubber, bitumen caoutchouc, paraffin wax
and fluid silicones, but does not dissolve cellulose esters. It is miscible
with other organic solvents.

Characteristics
Sp. gr. 0.782 (0.7784 at 20° C); n_D 1.4262; b. p. 80.74°C; s. p. 6.5° C;
v. p. 77; flash p. 37° F (3° C); visc. 0.94; dielec. const. 2; lat. ht. 86; sp.
ht. 0.497; cub. expn. 0.0011; evap. period 195 (toluene = 100); expl.
lim. 1.3–8.4%.

A.S.T.M. Spirits of Turpentine, D13–65

	Gum Spirits of Turpentine		Wood Turpentine					
			Steam Distilled		Sulphate		Destructively Distilled	
	Max.	Min.	Max.	Min.	Max.	Min.	Max.	Min.
Specific gravity 15.5/15.5 C.	0.875	0.860	0.875	0.860	0.875	0.860	0.865	0.850
Refractive index at 20 C., D line.	1.478	1.465	1.478	1.465	1.478	1.465	1.483	1.463
Residue, after polymerization with 38N H_2SO_4; Volume per cent.	2	–	2	–	2	–	2	–
Refractive index at 20° C.	–	1.500	–	1.500	–	1.500	–	1.480
Initial boiling point at 760 mm. pressure, °C.	160	150	160	150	160	150	157	150
Distilling below 170° C at 760 mm. pressure, per cent.	–	90	–	90	–	90	–	60
Distilling below 180° C at 760 mm. pressure per cent	–	–	–	–	–	–	–	90

It forms the following constant boiling mixtures:

Cyclohexane (%)		b. p. (°C)
45	benzene 55%	77.5
63	methyl alcohol 37%	54.2
70	ethyl alcohol 30%	64.9
77	isopropyl alcohol 23%	68.6
80	n-propyl alcohol 20%	74.3
86	isobutyl alcohol 14%	78.1
90	n-butyl alcohol 10%	79.8
92	water 8%	69.4
60	methyl ethyl ketone 40%	72.0
85	methyl glycol 15%	77.5
52	benzene 39% methyl glycol 9%	73

Methyl cyclohexane

Known also as hexahydrotoluene, it is produced by the hydrogenation of toluene and by the interaction of benzene and methane at high temperatures. It occurs naturally in Russian and Galician petroleum and in cracked petroleums.

Characteristics

When pure: sp. gr. 0.760 at 20° C; n_D 1.425; b. p. 101.2° C; m. p. −126° C; lat. ht. 77; sp. ht. 0.443; dielec. const. 2.1; visc. 0.64; flash p. 25° F (−4° C); evap. period 140 (toluene = 100).

Hydrogenated Solvent Naphtha

Hydrocarbons, having good solvent power for synthetic resins of the alkyl type, have been prepared by hydrogenating petroleum solvent naphthas. These products are miscible with blown castor oil, and more closely resemble the coal-tar solvent naphthas than do the original petroleum naphthas. The table on page 101 indicates the types of product available [5]:

Petroleum hydrocarbons

Many grades of petroleum hydrocarbon solvents are available: aromatic, cyclo-paraffinic, and aliphatic. They are all basically prepared from oil and are obtained during motor oil manufacture and result from

No	b.r. ($^\circ$C)	aniline p. ($^\circ$C)	flash p. (Tag) ($^\circ$F)	Dimethyl Sulphate value (%)	K-B value
1	93–135	11	below 60	24	55.2
2	135–185	−18	61	62	75.6
3	185–215	−23	135	87	77.3
4	215–238	−36	190	100	85.7

fractional distillation and purification as well as from chemical reaction, for example the conversion of aliphatics to aromatics. The actual constitution of each product needs to be understood in order to obtain the best possible value from its properties [13].

Petroleum Ethers (Ligroins)

These are the most volatile fractions and are generally of narrow boiling range, n_D 1.365–1.376 and high degree of purity, for example:

	1	2	3
sp. gr.	0.645	0.669	0.676
b. r. ($^\circ$C)	40–60	60–70	60–80
Aromatics	1%	4%	5%

Certain grades are used for pharmaceutical purposes, the following are standardized:

British Pharmacopaeia quality: sp. gr. 0.62–0.70; b. r. 95% between 60°–70° C.

U.S. Pharmacopaeia quality: sp. gr. 0.634–0.660. b. r. 35°–80° C. *Industrial Special Boiling-point Spirits.*

These include extraction benzines or naphthas, rubber solvents, lacquer diluents.

Extraction benzines vary widely in boiling range, specific gravity and aromatic hydrocarbon content; almost any boiling range between the limits of 30° and 160° C being available with specific gravities varying between 0.67 and 0.80, and aromatic contents between 6% and 45%.

The following grades are typical [18]:

	b. r. (° C)	sp. gr.
No. 1	35–115	0.675–0.695
No. 2	70– 95	0.700–0.720
No. 3	100–120	0.735–0.755
No. 4	40–150	0.705–0.730
No. 5	90–105	0.725–0.745
No. 6	140–160	0.770–0.790

They are good solvents for all types of oils with the exception of castor oil, in general those benzines with high aromatic content are the better solvents. K-B value about 30. Dil. rat. for cellulose acetate in acetone about 1.1 and for cellulose nitrate in butyl acetate about 1.2.

Rubber Solvents. Several types of petroleum solvents are used in processing rubber. For the preparation of solutions of rubber used for dipping and for the vulcanization processes a still lower boiling range is preferred such as 30°–100° C and for manufacture of moulded articles solvents having a boiler range of 100° to 160° C and sp. gr. 0.745 to 0.770 are preferred. More voltaile spirits having a boiling range of 70° C–120° C are used for work where rapid drying is required, while for cold. In general a high aromatic content leads to rubber solutions of high viscosity, while prolonged milling of the rubber reduces the viscosity of the solutions.

The following figures indicate the solvent characteristics of this class, K-B value about 37. Dil. rat. for cellulose nitrate in butyl acetate about 1.2; in acetone about 1.1; for cellulose acetate in acetone about 1.2

Lacquer Diluents. Hydrocarbons are non-solvents for cellulose esters but they are used to reduce the cost, to adjust viscosities and to permit the incorporation of certain resins into cellulose lacquers; for these purposes the aromatic hydrocarbons are superior as their precipitating effects on the cellulose esters are not so pronounced. Nevertheless, petroleum hydrocarbon distillates of high aromatic content can often be used successfully and certain grades may replace benzene, toluene and xylene if due allowance is made for the different dilution ratios; the following are typical distillates:

	sp. gr.	b. r. (° C)	Aromatics (%)
Benzene substitute	0.772	75–95	27
Toluene substitute	0.795	100–120	45
Xylene substitute	0.800	115–160	45

The dilution ratios for cellulose nitrate in butyl acetate are of the order 1.5 to 2.0 as compared with about 2.7 for the aromatic hydrocarbons. The dilution ratios for paraffin distillates of low aromatic (*e.g.* below 20%) content vary between 1 and 1.25.

White spirit

White Spirit is an extremely important petroleum product which may be regarded as the next product distilled over after the Special Boiling-point Spirits (SBPS).

Several grades are readily available — the normal material, (English and American Standard — see below), 115° F Flash White Spirit, 140° F Flash White Spirit, 180° F End-Point White Spirit, High Aromatic White Spirit and odourless White Spirit [6], (this is mainly synthetic—see H table iso-paraffins). Occasionally white spirit is still known as petroleum or mineral spirits, paint and varnish thinner, varnish makers' and painters' naphtha.

The different grades of White Spirit have been developed for various industrial uses over many years and find application in the manufacture of surface coatings where distillation range and aromatic content variations are used to control drying time, odour, and type of film formant. In dry cleaning, distillation range and aromatic content are important factors in cleaning efficiency together with after-odour of cloth; recovery of the White Spirit for re-circulation through the cleaning plant is another property closely related to the original specification. In polish manufacture odour, wax solvency (both natural and synthetic) as well as evaporation rate are all necessary characteristics — this latter is of especial interest for polishes used in tropical conditions. In the manufacture of emulsions or solutions of certain synthetic insecticides containing White Spirit chemical inertness of the latter is of great importance.

British Standard Specification, BS 245:1956 for white spirit requires: b. r. up to 155° C 10% max., below 195° C 90% min; max. b. p. 210° C; residue 155e C 10% max., below 195° C 90% min; max. p. p. 210° C; residue 0.005% max. flash p. 93° F min. (33.9° C); neutral; free from objectionable sulphur compounds.

ASTM Specification, D484—53 for stoddard solvent requires: a petroleum distillate, clear and free from suspended matter; flash p. 100° F min (38° C); b. r. at least 50% by 176° C, at least 90% by

190° C; max b. p. 410° C; max. residue 1.5%; neutral; free from objectionable odour.

ASTM Specification, D235–61 for Petroleum Spirits (mineral spirits), requires: flash p. 100° F min (38° C); b. r. at least 50% by 177° C; max. b. p. 210° C, residue neutral.

High aromatic petroleum hydrocarbons

A wide range of these materials is available – due to the economic necessity of finding uses for the by-products obtained during motor fuel manufacture. They are prepared in modern plants to high degrees of specification accuracy which embrace distillation range, aromatic content, specific gravity and freedom from trace impurities. Mention has already been made of the petroleum derived benzene, toluene, and xylene, so it may be reckoned that the present group embraces all other petroleum aromatics.

Proprietary names are frequently used for these products and there is no recognized general standard for any particular type of product except the manufacturers' own. Thus, an aromatic content of 88%, a distillation range of 162° – 180° C, a specific gravity of 0.877 and an n of 1.4985 would indicate a particular fraction from a specific manufacturer which, whilst consisting essentially of a mixture of three isomeric trimethyl benzenes, would be given a trade name by that manufacturer. Some adherence to old nomenclature persists in this sphere, e.g. High Aromatic Naphtha.

Generally, these petroleum aromatics have found use in the fields of surface coatings, inks, polishes, (solution and emulsion) insecticides, and extraction where their ready availability, consistency of quality, attractive price, low odour and accurately known distillation range can be used to tailor-make a product for almost any requirement.

The following table shows the basic properties of four such hydrocarbons. No. 4, although strictly outside this group, is included on account of its wide use.

	1	2	3	4
sp. gr.	0.877	0.859	0.883	0.716
b. r. (° C)	162–180	153–193	162–272	180–207
flash p. (Abel). (° F)	117	113	132	130
Aromatics % Vol.	98	84	81	largely aliphatic
K-B Value	90	75.8	81.3	26

Paint and Varnish Thinners. This is a general name for petroleum solvents used in the paint and varnish industry and covers a range of distillates with boiling ranges up to about $250°$ C. max.

V.M. and P. Naphtha. A distillate known in America as varnish makers' and painters' naphtha which is somewhat more volatile than white spirit. b. r. about $100° - 160°$ C; n_D 1.41; s. g. $0.75-0.77_{20}$; flash p. $20°-50°$ F $(-7°$ to $10°$ C).

Petroleum distillates are in general solvents for beeswax, carnauba wax, montan wax, Japan wax, paraffin wax [14], ceresin, spermaceti, bitumen, rubber, polyisobutylene, polybutadiene-isobutylene, fluid silicones, some melamine-formaldehyde resins, ester gum, dammar and elemi, their solvent powers generally increasing with increase in aromatic and naphthenic hydrocarbon content. They do not dissolve cellulose esters or ethers, shellac, pontianac, manila, copals, sandarac, chlorinated rubber, polystyrene, nylon, urea-formaldehyde resins. They attack polythene.

The characteristics of a few pure saturated aliphatic hydrocarbons are quoted below as a guide.

$b. p.$ °C	$sp. gr.$ (20°C)	$s. p.$ °C	$v. p.$	$lat. ht.$	$sp. ht.$	n_D
n-Pentane 36.0	0.626	−130	422	84	0.540	1.357
n-Hexane 68.6	0.659	−95	120	82	0.531	1.375
n-Heptane	98.2	0.684	−91	36	76	0.531
n-Octane 125.6	0.703	−57	10	71	0.518	1.398
n-Nonane 150.7	0.718	−54	3.2	−	0.523	1.405
n-Decane 173.7	0.731	−30	2.7	60	0.520	1.412

Pentane has found some application during the preparation of rigid plastic foams and 62–68 Hexane is employed in certain adhesives and as an inert carrier solvent for certain chemical reactions. Hexane is also used in extracting cotton deed, flax seed, and soya bean oils.

Styrene
(Vinyl benzene) 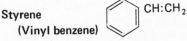 CH:CH$_2$

Styrene monomer has assumed some importance as a solvent in the field of unsaturated polyesters resins. It is present in the resin solution in amounts of generally around 30% by weight, but, unlike virtually every volatile solvent used in other coatings systems it is an essential part of the final film since it cross-links with the polyester resin to form an irreversible complex (after the addition of accelerator). This latter fact has possibly over-ridden the solvent role played by the styrene – a branch-chain aromatic usually prepared by dehydrogenating ethyl benzene.

Styrene should not be regarded as a petroleum spirit since it has a flash point of 88° F (31° C); the lower explosion limit is 1.1% and the upper 6.1% by volume of styrene in air and the vapours released from polyester solutions are heavier than air. It is important to ensure adequate ventilation in areas where styrene monomer may occur.

Hydroterpin

This substance is a product of the hydrogenation of turpentine, and has a similar mild odour. b. r. 180–195°C. sp. gr. 0.879 at 20° C. n_D 1.4773.

Tetrahydronaphthalene

1, 2, 3, 4-Tetrahydronaphthalene, known commercially under the trade-mark 'Tetralin,' is a product obtained by the partial catalytic hydrogenation of naphthalene, one ring being completely hydrogenated, the other remaining unchanged. This is effected by conducting the hydrogenation in the vapour phase. Tetrahydronaphthalene is an aromatic cycloaliphatic hydrocarbon.

It is a non-toxic liquid which oxidizes on exposure to air, like turpentine, forming substances which decompose violently on sudden heating [16]. The chief oxidation products of tetralin are α-tetralone, α-tetralol and γ-o-hydroxyphenylbutyric acid. Tetralone forms a hydroperoxide which changes to a 1,2-diketone and then to acidic substances [17].

Characteristics: sp. gr. 0.963–0.972; b. p. 200° C; b. r. 200–209° C; flash p. 172° F (78° C); s. p. −25° C; v. p. 0.18; n_D 1.540–1.547; lat. ht. 79; sp. ht. 0.40; K-B value 200+; ht. comb. 0.10150; Visc. 2.2; dielec. const. 2.66

Tetrahydronaphthalene oxidizes on heating in the presence of air and its boiling range widens.

It is a powerful solvent for oils, resins, waxes, rubber, polyvinyl chloride, coumarone, colophony, mastic, asphalt, most resinates, Albertols, and linoxyn. It does not dissolve cellulose esters, hard copals, shellac, bakerlites, and is not miscible with alcohol (unless anhydrous) or water.

Generally, tetrahydronaphthalene finds use as an additive to organic surface coatings and inks by virtue of its low evaporation speed — which enhances secondary flow even in lacquers based on cellulose derivatives. It also assists in cutting down sedimentation with metallic paints and in increasing the efficiency of solvent-based paint removers.

Decahydronaphthalene

Decahydronaphthalene, known commercially under the trade-mark names 'Dec' and 'Decalin,' [8] is produced by the complete hydrogenation of naphthalene or tetrahydronaphthalene, both in the liquid state, at temperatures ranging from 150° C. upwards, 190° C being the most favourable. It is a non-toxic liquid.

Characteristics: sp. gr. 0.873–0.886; b. r. 183–192° C; flash p. 135° F (57° C); m. p. −124° C; n_D 1.467–1.479; lat. ht. 71; sp. ht. 0.3874; evap. period 2.6 (turpentine = 1); Visc. 1.9; dielec. const. 2.1; auto-ignition temp 262°C

It is not such a powerful solvent as tetrahydronaphthalene and does not undergo atmospheric oxidation. Solvent for fats, waxes, rubber, dammar, mastic, manila. Non-solvent for cellulose esters, copal, kauri, linoxyn. Non-miscible with water or alcohol (unless anhydrous), but miscible with most organic solvents. Generally used as an additive to improve flow.

Carbon Disulphide. CS_2

Carbon disulphide is a colourless, highly volatile liquid which becomes yellow with age and under the influence of light. It has an ethereal odour which is more or less pleasant, according to the impurities present.

It is a solvent for sulphur, rubber, vegetable oils and ester gum; it gels ethyl and benzyl cellulose, but does not dissolve cellulose acetate or nitrate, vinyl resins or shellac. It attacks polyethylene, polyvinylchloride and methyl methacrylate.

British Standard Specification BS 622:1950 for carbon disulphide requires: sp. gr. 1.270–1.274 (1.265–1.269 at 20° C); b. r. up to

46.4° C 95% min; residue 0.01% max; neutral to methyl red; free from H_2S.

The pure substance has b. p. 46.25° C; m. p. −111.6° C; flash p. −8° F (−22° C); auto-ignition temperature 125° C; lat. ht. 84; sp. ht. 0.24; n_D 1.635; water dissolves 0.2%; dielec. const. 2.67; elec. cond. 3.7×10^{-3}; visc. 0.365; v. p. 298.

Carbon disulphide is dangerously inflammable, mere contact with a hot steam pipe or an electric lamp bulb is sufficient to cause the vapour to ignite by reason of its extremely low auto-ignition temperature. The explosive limits of mixtures with air lie between 1% and 50% by volume.

Carbon disulphide is narcotic in high concentration and is a severe chronic nerve poison.

The following azeotropic mixtures are known:

			b. p. (°C)	
carbon disulphide	14%	Methyl alcohol 86%	37.7	
carbon disulphide	8%	ethyl alcohol 92%	42.4	
carbon disulphide	7%	Isopropyl alcohol 93%	44.6	
Carbon disulphide	32.5%	Acetone 67.5%	39.2	
Carbon disulphide	15%	Methyl ethyl ketone 85%		45.9
Carbon disulphide	36.5%	Ethyl formate 63.5%	39.4	
Carbon disulphide	30%	Methyl acetate 70%	40.2	
Carbon disulphide	99%	Ethyl ether 1%	34.5	

Nitroparaffins

These solvents [8, 10] are made by reacting the lower paraffin hydrocarbons with nitric acid under high pressure at temperatures up to 550° C; a continuous process [15] is used involving a very short time of reaction.

The chief products of the nitration of propane are nitromethane, nitroethane, 1-nitropropane and 2-nitropropane together with ketones, aldehyde and olefine by-products, the last being rehydrogenated and returned to the process.

The table on page 109 gives the physical properties of the more important nitroparaffins.

They are colourless, non-hygroscopic liquids of mild odour but there are indications that they are toxic [11] about 0.1% in air may be assumed to be dangerous, causing irritation of the central nervous system and an increase of the nitrate and nitrite content of the blood. The toxicity of the nitro-paraffins increases with molecular size [12].

	Nitro-methane	Nitro-ethane	1-Nitro-propane	2-Nitro-propane
sp. gr. at 20° C.	1.139	1.052	1.003	0.992
n	1.3818	1.3916	1.4015	1.3941
b. p. ($^\circ$C)	101.2	114.0	131.6	120.3
m. p. ($^\circ$ C)	-29	-90	-108	-93
v. p.	27.8	15.6	7.5	12.9
evap. range (Butyl acetate - 100)	180	145	100	124
s. ten.	37	31	30	30
Water dissolves % at 20° C	9.5	4.5	1.4	1.7
Dissolves water % at 20° C	2.2	0.9	0.05	0.6
dielec. const.	39	–	–	–
dil. ratio (toluene)	–	–	1.2	1.3

The nitropropanes are solvents for cellulose nitrate, cellulose acetate-butyrate, ethyl cellulose, benzyl cellulose, ester gum, coumarone, glyptals, buna N, chemigum, hycar OR, butacites, butadiene-acrylonitrile resins, polystyrene-maleic anhydride copolymers, rubber chloride, polyvinyl acetate, aceto-chloride and acetals, vegetable and castor oils, and low boiling hydrocarbons. In the presence of alcohol they dissolve cellulose acetate and aceto-propionate, aceto-butyrate, kauri, manila, pontianac and formvar. They partly dissolve polyvinyl-chloride, dammar, kauri, colophony. They do not dissolve cellulose triacetate, methyl cellulose, polyacrylonitrile, tars, pitches, waxes, casein, gelatin, zein, congo, shellac, rubber, buna S or high boiling hydrocarbons.

Chloro-nitroparaffins

Two members of this new class of solvent are now available in small amounts [8, 10]. They are colourless liquids of mild but penetrating odour and are more powerful solvents than their parent nitroparaffins, in particular for many synthetic rubbers, they prevent the gelation of rubber solutions and are insoluble in water. They are miscible with most organic solvents including the lower alcohols, glycols, esters, ethers, petroleum hydrocarbons, mineral and vegetable oils, fats and waxes.

1-chloro-1-nitropropane has sp. gr. 1.209 at 20° C; b. r. 139–143° C; flash p. 144° F (52° C); n_D 1.430.

2-chloro-2-nitropropane has sp. gr. 1.193 at 20° C; b. r. 129°–134° C; flash p. 47° C; n_D 1.425.

References

[1] *Bull. Soc. Chim.,* Paris, IV., **39**, 873, 1926.
[2] Shell. Chem London, Imperial Chemical Industries Ltd.
[3] Esso Petroleum London.
[4] Howards and Son Ltd., Ilford, London, E., Imperial Chemical Industries Ltd.
[5] Sweeney and Tilton, *Ind. Eng. Chem.,* 1934, p. 693. Standard Oil Co., Lousiana, U.S.A., W. J. Hough & Co., 2545 W. Congress Street, Chicago, U.S.A.
[6] Carles Capel-Leonard London E.
[7] Brennstoffchemie, 1933, p. 106.
[8] Imperial Chemical Industries, London. Brit. Pat. 578044, 575733, 576129, 589480.
[9] Lush. *Chem. and Ind.,* 1927, p. 454T.
[10] Commercial Solvents Corp., N.Y., U.S.A. *Chem. and Met. Eng.,* Sept., 1942, p. 129. *Ind. Eng. Chem.,* 1943, p. 1026; 1942, p. 1091. *Chem. Industries,* 1943, p. 664.
[11] *J. Ind. Hyg. Toxicology,* Oct., 1940.
[12] "Paint Manufacture," Jan., 1941, p. 4.
[13] *Ind. Eng. Chem,* 1938, p. 807.
[14] *J. Inst. Pet.,* 1941, p. 369.
[15] Hercules Powder Co., Wilmington, Delaware, U.S.A.
[16] Private communication. Dr. H. J. Stern, London, W.6.
[17] A. Robertson and W. A. Waters. *J.C.S.,* 1948, p. 1574.
[18] U.S.P. 2511454–1950.

Alcohols and their ethers

Methyl Alcohol. CH_3 . OH

Methyl alcohol or methanol is an important solvent, the manufacture of which has attracted very considerable attention during the last decade.

Formerly it was manufactured entirely by the distillation of wood along with acetic acid, acetone (q.v.) and other substances (cf. Wood spirit). Methanol made by modern synthetic methods from water-gas or natural gas (predominantly methane) has now displaced the natural product.

The most important synthetic method is that originally discovered by Sabatier in 1905 [1], who effected the catalytic combination of carbon monoxide and hydrogen [2]. In 1913 the Badische Anilin und Soda Fabrik patented [3] the use of a wide range of catalysts for promoting the conversion of the constituents of water-gas into a complex mixture of methyl and other alcohols, aldehydes and ketones. A variety of patents quickly followed, but perhaps the most out-standing work is that of Patart in France, who used as raw materials either carbon monoxide and hydrogen [4] or methane and oxygen [5]. Thus one volume of carbon monoxide with one and a half to two volumes of hydrogen at 150 to 200 atmospheres pressure, passed over zinc oxide at a temperature of 400° to 450° C, gives a liquid product which consists mainly of methyl alcohol and water with traces of aldehydes and acetone. Similarly, water-gas and hydrogen at 500 atomspheres pressure give methyl alcohol in a yield of 80%, together with traces of higher alcohols, acids, water and ammonia but no acetone, 300 cubic metres of water-gas yielding 100 litres of methyl alcohol. Patart also found that methane mixed with one-half its volume of oxygen, when passed over coke at a temperature of 800° to 1,000° C, yields a gas consisting of two volumes of hydrogen with one volume of carbon monoxide, this mixture on further catalytic treatment as above, yielding a liquid consisting mainly of methyl alcohol [6]. Further, coal gas or coke-oven gas, passed over zinc oxide or chromium sesquioxide at a temperature of 300° and under a pressure of 150 to 200 atmospheres, similarly yields a liquid consisting of methyl alcohol and hydrocarbons [7]; the hydrocarbons, on further catalytic treatment, give higher alcohols in considerable yield. The addition of alkali hydroxides or carbonates to the usual catalysts causes an increase in the proportion of higher alcohols [8]. This process, of which there are many variations regarding details and catalysts, has been developed to one of considerable industrial importance for the manufacture also of higher alcohols, particularly isobutyl alcohol.

It has also been proposed to manufacture methyl alcohol from methane by chlorination [9] and treatment of the resulting methyl chloride with alkali acetates [10]. This process is closely linked with that for the production of methylene dichloride and chloroform (q.v.)

Methyl alcohol is known in America as Columbus spirit, although methanol is the term now usually applied there.

Characteristics (Pure): sp. gr. 0.796; b.p. 64.7° C; flash p. 32° F (0° C); ign. p. 0° C; m.p. −97° C; v.p. 95; n_D 1.329; lat. ht. 263; sp. ht. 0.597; ht. comb. 5334; cub. expn. 0.00118; visc. 0.61; s. ten. 23; therm. cond. 0.0005 at 30° C; elec. cond. 4.4×10^{-7}; dielec. const. 31.

British Standard Specification, BS 506:1966 for methyl alcohol requires: sp. gr. 0.799 max, (0.796_{20}); b.r. 64.4−65.5 (95%); acidity 0.004% max. as acetic acid; residue 0.01% max; aldehydes and ketones 0.06% max. as acetone; sulphur (total) 0.001% max.; no opalescence with 19 vols. of water.

ASTM Specification, D1152−58: for Methyl alcohol, requires: sp. gr. max. 0.7928; b.r. 64.0°−65.6° C; max non-volatile material 0.005 g/100 ml; not more than 0.15% H_2O by weight, acidity not more than 0.003% by weight, acetone not more than 0.003% by weight.

Anhydrous methyl alcohol is clearly miscible with benzene in all proportions. It does not form a constant-boiling mixture with water, but it does so with many organic solvents. The following are known:

Methyl alcohol (%)	(%)	b. p. (°F)
15	Acetone 85	55.9
19	Methyl acetate 81	54.0
40	Benzene 60	58.3
62	n-Heptane 38	60.5
27	n-Hexane 73	50.0
21	Carbon tetrachloride 79	55.7
36	Trichlorethylene 64	60.2
37	Cyclohexane 63	54.2
16	Ethyl formate 84	51.0
47	Ethyl acetate 53	74.0
14	Carbon disulphide 86	37.6
12	Chloroform 88	54.0

Pure methyl alcohol dissolves neither cellulose acetate nor high-viscosity nitrate, but the presence of a relatively small proportion of acetone renders it a potent solvent for the nitrate and, to a less degree, for the acetate. It dissolves '½ sec.' cellulose nitrate fairly readily. It is a solvent for ethyl cellulose, polyvinyl acetate, colophony, shellac, benzyl abietate, soft bakelites, castor oil, zein, 'gammexane', and is miscible with aromatic hydrocarbons. It partially dissolves mastic and soft copals. It is non-miscible with linseed oil and with petroleum hydrocarbons. It does not dissolve polyvinylchloride or chloracetate.

Ethyl Alcohol. C_2H_5 . OH

Ethyl alcohol or ethanol is mainly produced by two methods, the fermentation of 'black strap' molasses and by absorbing ethylene from petroleum in sulphuric acid and hydrolysing. It has been very largely used in conjunction with ethyl ether as a solvent for cellulose nitrate in the manufacture of celluloid. This mixture is now seldom used for lacquer work. Alcohol, by itself, is not generally a solvent for cellulose nitrate or acetate, although anhydrous alcohol will dissolve some of the lower nitrated forms of cellulose. The solvent properties of alcohol rapidly decrease with increasing proportions of water; alcohol containing 5% of water is practically a non-solvent for all forms of cellulose nitrate.

Anhydrous alcohol dissolves colophony, sandarac, kauri, pontianac, manila, elemi, mastic, thus, dammar, shellac, gum, camphor and heat-treated congo and zanzibar; an addition of 20 to 30% of benzene renders it a solvent for benzyl cellulose, ester gum, and coumarone. It dissolves ethyl cellulose and solutions up to 16% by volume can readily be made; such solutions may be clarified by the addition of a small proportion of toluene if necessary. It is a solvent for certain synthetic resins such as the soft bakelites, polyvinyl acetate, low viscosity silicones, acetaldehyde resin, acrolein resin, furfural resin, cyclohexanone resin, and cyclohexanone-formaldehyde resin. It is miscible with castor oil and hydrocarbons.

Pure anhydrous alcohol has the following characteristics: sp. gr. 0.7937; p. p. 78.3° C; flashp. 57° F (14° C); ign. p. 371–427° C; m.p. −115° C; v.p. 44; $n_D1.3619$; lat. ht. 209; sp. ht.0.588; ht. comb.7130; cub. expn. 0.00108; visc. 1.2; s. ten. 22; therm. cond. 0.00039 at 20° C; elec. cond. 1.4×10^{-9}; dielec. const. 26.

British Standard Specification, BS 507:1966 for Ethanol, requires: sp. gr. at least $^{20}/_{20}$ 0.83014 equivalent to 58 degrees overproof (85.84%

by weight, 90.14 by volume); max. residue 100 ppm; max. acid 50 ppm; not more than 0.10% aldehydes and ketones, no turbidity with 19 volumes of distilled water.

Methylated spirit varies in composition. A form most suitable for lacquer manufacture consists essentially of the British Standard alcohol quoted above denatured with 5% of methyl alcohol and is more usually termed *industrial alcohol.*

The following azeotropic mixtures are known:

Ethyl alcohol (%)	(%)	b.p.($^{\circ}$C)
95.57	Water 4.43	78.15
32.37	Benzene 67.63	68.24
n-Propyl Alcohol. CH3 . CH2 . CH2 . OH		

Propyl alcohol or propanol is one of the chief constituents of fusel oil, sp. gr. at least 0.83014 equivalent to 58 degrees overproof (85.84% by weight, 90.14 by volume)

n-Propyl Alcohol. CH_3 . CH_2 . CH_2 . OH

Propyl alcohol or propanol is one of the chief constituents of fusel oil, from which it is isolated in a state of high purity. In its properties it is similar to isopropyl alcohol, to which it is superior as a lacquer diluent on account of its higher boiling-point, being therefore less likely to cause chilling. It is miscible with all the usual solvents and with water in all proportions.

Characteristics (Pure): sp. gr. 0.8053_{20}, 0.8016_{25}; b. p. 97.2° C; flash p. 72° F (22° C); m. p. -127° C; n 1.3862; lat. ht. 163; sp. ht. 0.586_{25}; visc. 2.29, 2.015_{25}; s. ten. 23.9, 23.5_{25}; therm. cond. 0.00038_{30}; elec. cond. 9×10^{-9}; dielec. const. 22.

It forms a number of constant-boiling mixtures, of which the following are of interest:

n-Propyl alcohol (%)	(%)	b. p. ($^{\circ}$C)
71.7	Water 28.3	87.7
20	Cyclohexane 80	74.3
4	n-Hexane 96	65.6
38	Heptane 62	84.8
17	Benzene 83	77.1
49	Toluene 51	92.6
74	n-Octane 26	95.0
11.5	Carbon tetrachloride 88.5	72.8

n-Propyl alcohol (%)	(%)	b. p. ($^{\circ}C$)
54	Tetrachlorethylene 46	94.0
54	Perchlorethylene 46	94.0
80	Monochlorbenzene 20	96.5
57	Diethyl ketone 43	94.9
40	n-Propyl acetate 60	94.2
9.8	n-Propyl formate 90.2	80.6
51	Ethyl propionate 49	93.4

It is a solvent for colophony, soft copals, shellac, soft bakelites, ester gum, benzyl abietate, urea-formaldehyde resins, castor oil. When hot it dissolves cottonseed and linseed oils. It does not dissolve cellulose esters, but has a solvent action on some cellulose ethers. It dissolves 5% w/v of pure gammexane at 20° C.

Isopropyl Alcohol   CH₃\
                              CH.OH
                         CH₃/

Isopropyl alcohol (Propan-2-ol), known also as isopropanol and secondary propyl alcohol, is widely used as a substitute for ethyl alcohol, uses including, surface coatings, nitrocellulose damping, cosmetic preparations and pharmaceutical manufacture. It is not a usual constituent of normal fusel oils, but is manufactured either from acetone or from the propylene arising from the treatment of petroleums, this latter source being now predominant.

There are several processes for the preparation of isopropyl alcohol from acetone, all depending on the catalytic addition of hydrogen by passing acetone and hydrogen gas over heated nickel catalyst. The processes vary mainly in the degree of pressure, time of contact, temperature, and means of supporting the nickel catalyst [11]. Thus acetone in the vapour state is readily hydrogenated at 115° to 125° C, whilst under pressures sufficiently high to keep the acetone liquid temperatures between 250° and 300° C are used. The catalyst can be either of the 'rigid' or the 'non-rigid' type, the former being in the form of nickel turnings which have been activated first by anodic oxidation in sodium carbonate solution [33], followed by reduction in hydrogen at temperatures ranging from about 180°–300° C. The non-rigid type of catalyst consists of kieselguhr or some similar 'support' upon which hydrated nickel carbonate has been precipitated, followed by roasting and reduction in hydrogen at 250° C [12].

Granular nickel catalyst has also been used, the acetone being distilled over it in an atmosphere of hydrogen at 100 atomspheres pressure.

The manufacture of isopropyl alcohol from petroleum is conducted briefly as follows. Petroleum fractions of high-boiling range are 'cracked' by treatment at a high temperature, this cracking process giving rise to petroleums of low-boiling range and to a large quantity of gas. The gas has an olefine content of about 10 to 15%, and contains ethylene, propylene, butylenes, and amylenes; it is first freed from sulphur compounds, and then passed up through a series of absorption towers, down which are sprayed sulphuric acid solutions of graduated strengths, so as first to absorb the amylenes and butylenes, then the propylene and finally the ethylene [13]. The acid and gas are also circulated counter-current-wise so that the fresh gas meets the diluted acid, and *vice versa*. After absorption of the olefines the saturated acid is mixed with water, thus causing higher alcohols and polymerised olefines to separate out as an oil layer; the clear acid solution is then steam-distilled and crude isopropyl alcohol obtained (cf. secondary butyl alcohol and synthetic amyl alcohol). The crude alcohol is subjected to fractional distillation and is obtained in a high state of purity.

The pure alcohol has an odour slightly stronger than that of ethyl alcohol, and its taste is slightly bitter. It is a stable colourless liquid, hygroscopic, miscible with water in all proportions.

Characteristics: sp. gr. $^{20}/_{20}$ 0.7863; b. p. 82.4° C; flash p. 12° C; m. p. −88.5° C; v. p. 33; n_D 1.3772; lat. ht. of fusion 1293 Kcal/mol; sp. ht. at 20° C 0.599 cal/gm; ht. comb. 7970; cub. expn. 0.00107; visc. 2.2; s. ten. 22; elec. cond. 35 × 10^{-7}; dielec. const. at 25° C 18.3.

British Standard Specification, BS 1595:1965 for isopropyl alcohol, requires: sp. gr. 0.789−0.791 (0.786−0.788$_{20}$); b. r. 81.5°−83.0°C; max. residue 20 ppm; not alkaline to phenolphthalein and max. 20 ppm by weight of acid; not more than 0.50% H_2O; not more than 0.10% aldehydes and ketones; no turbidity with 19 volumes of distilled H_2O. *ASTM Specification*, D770−64 for isopropyl alcohol, requires:s sp. gr. 0.785−0.790 at 20° C; b. r. within 1.5° C and including 82.3° C; max.

residue 0.005 g/100 ml; max. acidity 0.002% by weight; no turbidity with 10 volumes distilled H_2O at 25° C; max. H_2O 0.2% by weight.

Various grades of isopropanol are available: commercially pure, 87% isopropanol 13% water azeotrope, and a similar azeotrope but with a small addition of methanol.

It forms the following constant-boiling mixtures:

Isopropyl alcohol (%)	(%)	b. p. (°C)
87.7	Water 12.3	80.4
33	Cyclohexane 67	68.6
33	Benzene 67	71.9
22	n-Hexane 78	61.0
69	Toluene 31	80.6
8	Carbon disulphide 92	44.6
30	Methyl ethyl ketone 70	77.3
23	Ethyl acetate 77	74.8
52	Isopropyl acetate 48	80.1
10.0	Isopropyl ether 90.0	64.9
4.2	Chloroform 95.8	60.8
14	Carbon tetrachloride 86	67.0
45	Ethylene dichloride 55	74
28	Trichlorethylene 72	74
81	Tetrachlorethylene 19	81.7

And with 7.5% of water and 73.8% of benzene boiling at 66.5° C.

The azeotrope with water b.p. 80.4° C has m.p. −50° C.

Pure isopropyl alcohol is not a solvent for cellulose esters, but has distinct latent solvent properties, the presence of a relatively small quantity of an ester rendering it a solvent for cellulose nitrate. It dissolves cyclohexanone-formaldehyde resin, colophony, shellac, kauri, sandarac, manila and low viscosity silicones; it partly dissolves mastic, elemi, dammar and soft copal, it does not dissolve polyvinyl chloride, acetate or chloracetate.

It has the advantage over industrial alcohol in that it is anhydrous and therefore less likely to cause water blush. The fact that no methylation is legally required is also in its favour, since methylating agents can affect the strength and colour of cellulose films.

n-Butyl Alcohol. $CH_3 . CH_2 . CH_2 . CH_2 . OH$

n-Butyl alcohol, known also as butanol, is of extreme industrial importance, since it serves as a raw material for the manufacture of n-butyl acetate, the most widely used of the cellulose-nitrate solvents.

n-Butyl alcohol is mainly manufactured by two processes; a large part of the butyl alcohol of commerce is produced by the fermentation process developed by Strange, Graham, Fernbach, Weizmann and others; improved processes are now in operation using *Clostridium butylicum* for the fermentation stage [14]. The Weizmann process consists in preparing a wash from maize flour by heating with water to 130°–140° C under two to three atmospheres pressure for three to four hours; inoculating the cooled, sterile wash with a specially prepared culture, fermenting for about forty-eight hours, and distilling the fermented wash, thus obtaining about 20 lb of crude solvent for every hundredweight of maize flour. This crude solvent consists mainly of acetone, and n-butyl alcohol with small quantities of ethyl alcohol, primary amyl alcohol, n-hexyl alcohol and their esters of acetic, n-butyric, caprylic and caproic acids. The n-butyl alcohol is separated from the wash by means of a continuous Coffey still, from which it is obtained with a dissolved water-content of about 9%. The anhydrous alcohol results on fractional distillation, the water coming over in the first fractions as a binary mixture, containing 37% of water.

Butyl alcohol has also been manufactured in very large quantities from acetaldehyde, itself obtained either from acetylene by hydration in the presence of a mercury salt or from alcohol by limited oxidation or dehydrogenation. In this process acetaldehyde is converted *via* aldol to crotonaldehyde, which is hydrogenated to n-butyl alcohol.

Butyl alcohols have also been made from water gas and hydrogen [15] or from coal gas and coke-oven gas by catalytic treatment under pressure, and at high temperatures (*vide* methyl alcohol) [16]. Ethyl alcohol circulated over magnesium, calcium, barium or manganese oxide at 400°–500° C forms butyl alcohol, esters and acetals, in 20 to 30% yield. The use of a catalyst consisting of 60 to 80% of magnesium oxide and 20 to 40% of copper oxide at a temperature between 270° and 325° C and under 200 to 300 atmospheres pressure has also been proposed [17].

Contrary to statements which have been made, n-butyl alcohol is not a normal constituent of fusel oils. n-Butyl and iso-butyl alcohols are manufactured on a large scale from propylene by the OXO carbonylation process.

Characteristics (Pure): sp. gr. 0.8102_{20}, 0.8066_{25}; b. p. $117.7°$ C; flash p. $116°$ F ($47°$ C); m. p. $-90°$ C; v. p. 4.3; n_D 1.3992; lat. ht. 141; sp. ht. 0.565; ht. comb. 8626; cub. expn. 0.00095; visc. 2.98, 2.60_{25}; s. ten. 24.8, 24.35_{25}; therm. cond. 0.00037 at $20°$ C; elec. cond. 9×10^{-9}; dielec. const. 18.

British Standard Specification, BS 508:1966 for butyl alcohol, requires: clear and free from suspended matter, essentially butan-1-ol; sp. gr. 0.813–0.815 (0.810–0.812 at 20° C); b. r. 116.0°–119.0° C; flash p. not lower than 90° F (approx 32° C); not more than 0.25% by mass H_2O; max. residue 50 ppm; max. acidity 50 ppm; not more than 0.20% by mass of aldehydes and ketones.

ASTM Specification, D304–58 for butyl alcohol, requires: sp. gr. 0,810–0.813 at 20° C; b. r. within 1.5° C including 117.7° C; max. residue 0.005 g/100 ml; max. H_2O 0.1% by weight; max. acidity 0.005% by weight.

The mutual solubilities of water and butyl alcohol are as follows:

Temperature (°C)	Solubility of butyl alcohol in water (%) by weight	Solubility of water in butyl alcohol
0	10.5	–
10	8.9	19.7
20	7.8	20.0
30	7.1	20.6
40	6.6	21.4
50	6.5	22.4
60	6.5	23.6
70	6.7	25.2
80	6.9	26.4
90	7.9	30.1
100	9.2	33.8
110	10.5	37.6
120	16.1	46.0

n-Butyl alcohol forms the following binary constant-boiling mixtures:

n-Butyl alcohol (%)	(%)	b. p. (°C)
62	Water 38	92.4
47	n-Butyl acetate 53	117.2
23.7	n-Butyl formate 76.3	105.8
40	Methyl iso-valerate 60	113.5
13	Heptane 87	90.0
4	Cyclohexane 96	79.8
27	Toluene 73	105.7
29	Tetrachlorethylene 71	109
32	Perchlorethylene 68	110
17	Ethyl isobutyrate 83	109.2

Two ternary mixtures are of interest, viz.: Butyl alcohol 10%; butyl formate 68.7%; water 21.3%; b.p. 83.6° C. Butyl alcohol 27.4%; butyl acetate 35.3%; water 37.3%; b.p. 89.4° C.

Butyl alcohol possesses the unique property of being a solvent for hard copals; it dissolves kauri, congo, manila, dammar, sandarac, elemi, shellac, ester gum, low viscosity silicones, cyclohexanone-formaldehyde, melamine-formaldehyde and urea-formaldehyde resins, benzyl abietate, polyvinyl acetate, calcium, zinc and manganese resinates, castor and linseed oils. It partly dissolves mastic and coumarone; it is miscible with hydrocarbons. It does not dissolve cellulose esters or ethers, polyvinyl chloride, polyvinyl chloride-acetate, or rubber chloride. The addition of a small quantity (3%) will bring about homogeneous mixing of methylated spirit with petroleum hydrocarbons. Ethyl cellulose dissolves in mixtures of butyl alcohol and xylene.

Isobutyl Alcohol. $(CH_3)_2 CH . CH_2 . OH$

Known also as isopropyl carbinol and 2-methylpropanol-1 the generally accepted industrial name is isobutanol. It is produced as a by-product during the synthesis of methanol and the oxidation of natural hydrocarbon gases. visc. 3.82, 3.24

It has found some application in the manufacture of cellulose nitrate lacquers, in stoving and catalysed finishes containing amino-resins, and in wash primers. It also behaves satisfactorily in phenol-formaldehyde coatings, PVC emulsions, epoxy primers, and any coating where a medium boiling alcohol is necessary. Generally, its characteristic behaviour lies between that of n-butanol and s-butanol but is usually cheaper than the former.

Characteristics: sp. gr. 0.8057, 0.8030_{20}, 0.7994_{25}; b. p. 108° C; flash p. 77° F (22° C); m. p. −108° C; v. p. 8.8; n_D 1.3962; lat. ht. 138; sp. ht. 0.603; cub. expn. 0.00095; visc. 3.82, 3.24_{25}; s. ten. 23.0, 22.55_{25}; elect. cond. 8×10^{-8}; dielec. const. 18; water dissolves 10% at 15° C; dissolves 15% water at 15° C.

The following constant-boiling mixtures are known:

Isobutyl alcohol (%)	(%)	b.p.(°C)
67	Water 33	90.0
14	Cyclohexane 86	78.1
9	Benzene 91	79.8
44	Toluene 56	101.1
9	M.I.B.K. 91	107.5

Secondary Butyl Alcohol

$$\begin{array}{c} OH \\ CH_3-CH-CH_3-CH_3 \end{array}$$

Known also as methyl ethyl carbinol and as butan-2-ol. It is made by hydrogenating methylethyl ketone, and also from the butylenes arising from the cracking of petroleum hydrocarbons by processes similar to those used for the manufacture of isopropyl alcohol (q.v.). It is widely used for many solvent applications, particularly as a chemical intermediate for the automobile and mining industries, but is available technically. The industrial product is a mixture of the two optical isomers.

Characteristics: sp. gr. 0.807–0.809 (20/20); b. p. 99.5° C; flash p. 75° F (24° C); m. p $-89°$ C; v. p. 12; n_D1.397; lat. ht. 134; sp. ht. 0.67; cub. expn.0.00097; s. ten. 23; solubility in water 22%.

It is a solvent for rosin, ester gum, penta ester gum, shellac, kauri, sandarac, mastic, elemi, manila, copal, and dewaxed dammear. It will also dissolve castor oil, linseed oil, aromatic hydrocarbons, ethyl cellulose, alcohol soluble phenolics, sulphenomide resin and polyvinyl butyral. It exhibits latent solvent power for cellulose nitrate but is not a solvent for cellulose acetate, triacetate or acetate butyrate, coumarone indene resins, some alkyd and phenolic resins, polyvinyl chloride and its copolymers, acrylic resins, chlorinated rubber, chlorinated PVC, chlorinated polyethylene or polystyrene and its copolymers.

It finds use as a medium boiling latent solvent in a wide variety of coatings, as a main solvent in insecticides and as a component of brake fluids, cutting oils, and industrial cleaning compounds. Its odour is reminiscent of peppermint. It forms a constant-boiling mixture containing 27.3% of water boiling at 87.5° C, and also one containing 13.7% sec. butyl acetate boiling at 99.6° C, many other azeotropes are known.

British Standard Specification, BS 1993:1953 for secondary butyl alcohol requires: sp. gr. 0.811–0.813 (0.807–0.809 at 20° C); b. r. 95% between 98.0° C and 101.0° C; max. residue 0.002%; max. H_2O 0.50%; max. acidity 0.003%;

ASTM Specification, D1007–58 (1965) for secondary butyl alcohol requires: sp. gr. 0.807–0.809 at 20° C; b. r. 98.0°–101.0° C; max. residue 0.005 g/100 ml; max. H_2O 0.5% by weight; max. acidity 0.002% by weight.

Tertiary Butyl Alcohol (CH$_3$)$_3$. C . OH

Known also as 2-methylpropan-2-ol it is occasionally used on account of its miscibility with water and all the usual solvents. The pure alcohol is a hygroscopic crystalline solid at ordinary temperature, it has an odour resembling camphor. It is synthesised on a large scale from cracked petroleum gas. It is used as a chemical intermediate and is also of use in the Pharmacutical industries as a solvent.

Characteristics: sp. gr. 0.783$_{25}$; b. p. 82.4° C; flash p. 48° F (9° C); m. p. 25° C; v. p 31; n_D 1.388; lat. ht. 131; cub. expn. 0.00135; visc. 3.35$_{30}$; s. ten. 20.

Amyl Alcohol

There are eight isomeric amyl alcohols, and three of these eight exist each in two modifications, owing to the presence of asymmetric carbon atoms.

The amyl alcohol of commerce obtained from fusel oil is a mixture of two of the eight, viz., primary isoamyl alcohol and active amyl alcohol, known as isobutyl carbinol and s-butyl carbinol, also as 3-methylbutanol and 2-methylbutanol respectively, their formulae being:

$$CH_3\!\!\diagdown\!\!{}_{\diagup}\!\!CH.CH_2.CH_2:OH \quad \text{and} \quad CH_3.CH_2\!\!\diagdown\!\!{}_{\diagup}\!\!CH.CH_2.OH$$
$$CH_3 \qquad\qquad\qquad\qquad\qquad CH_3$$

the latter having an asymmetric carbon atom.

Synthetic amyl alcohol is also available, being produced in America and sold under the trade name 'Pentasol' [18]. 'Pentasol' consists of a mixture of five of the eight isomeric amyl alcohols, and is manufactured from petroleum. The process begins with the isolation of a mixture of n-pentane and isopentane from light petroleum by fractional distillation, the fraction boiling between 28° and 39° C being taken.

This pentane fraction is dehydrated by means of by-product hydrochloric acid gas and then mixed in vapour state with chlorine gas in a modified venturi mixing chamber, then passed through a pipe-still reactor at 180° to 300° C, rapidly cooled and fractionally distilled to separate the chloropentanes from pentane and hydrochloric acid. The crude mixture is again fractionally distilled for the purpose of removing polychlorinated pentanes from the mono-chloropentanes. The mono-chloropentane mixture thus obtained has the following approximate composition:

	(%)
1-Chloropentane	24
2-Chloropentane	8
3-Chloropentane	18
2-Methyl-4-chlorobutane	15
2-Methyl-2-chlorobutane	30
2-Methyl-2-chlorobutane	5

The amyl alcohols are produced from this mixture by hydrolysis with sodium oleate solution in the presence of a catalyst. The process is made continuous by using two hydrolysing digestors in parallel, from one of which the products of the reaction pass to a series of fractionating columns while the hydrolysis is proceeding in the other digestor. The fractionating columns separate the amyl alcohols from the amylenes which are also produced and the unchanged amyl chlorides, these two latter being returned to the process. The amyl alcohol mixture finally obtained has the following approximate composition:

	(%)
Pentan-1-ol	26
Pentan-2-ol	8
Pentan-3-ol	8
3-Methylbutan-1-ol	16
2-Methylbutan-1-ol	32

There are also obtained tertiary amyl alcohol and diamyl ethers.

Amyl acetate is prepared by a continuous process in which the amyl alcohols with glacial acetic acid are fed into a fractionating still containing sulphuric acid, amyl acetate and water being distilled off.

In another similar process [19] the pentanes are treated with chlorine in an apparatus designed to regulate and limit the violence of the reaction, which, being exothermic, may proceed explosively. The apparatus consists of a lead-free glass tube of wide bore placed vertically; near the bottom of the tube is a screen, upon which rests activated carbon to a depth of about 3 inches; through the bottom of the tube, and through the screen, passes a small nozzle, by means of which the chlorine is injected into the funnel-shaped bottom end of a narrow glass tube placed concentrically within the larger one, and extending about one half of the way up. The pentane is placed in the apparatus, so as to fill it to the extent of about three-quarters, and is exposed to ultraviolet light, which aids the reaction. The passage of the chlorine through the jet at the bottom causes both the pentane and the catalyst to circulate in intimate contact, and the reaction proceeds smoothly; any vapour which is given off is returned to the apparatus by means of a refrigerating condenser, the hydrochloric acid gas produced in the reaction passing away through this condenser. In order to prevent or limit the formation of polychloro derivatives, the chlorination is discontinued when 20 to 25% of the pentane has been chlorinated; the product is then neutralised and the five amyl chlorides, produced in the reaction, fractionally distilled out.

The purified amyl chloride mixture produced by chlorinating in the liquid phase is of somewhat different composition to that obtained by vapour phase chlorination and consists approximately of:

	(%)
1-Chloropentane	26
2-Chloropentane	18
3-Chloropentane	8
2-Methyl-4-chlorobutane	16
2-Methyl-2-chlorobutane	32

The next stage in the process is the conversion of these chlorides to the corresponding acetates, and is conducted in autoclaves under a pressure of 200 to 205 lb per square inch, and at a temperature of $200°-230°$ C, the reagent being a mixture of powdered sodium acetate and carbon. The heating is maintained for eight hours, and the course of the reaction can be followed by noting the gradual drop in pressure as the highly volatile amyl chlorides are converted into the acetates. When the reaction is complete the crude amyl acetate is distilled out, neutralised and redistilled. The mixture of amyl acetates thus obtained has approximately the following composition:

	b. p. ($°$ C)	(%)
Pentanol-1	138	4
2-Methylbutan-1-ol	128	4
3-Methylbutan-1-ol	131	2
Pentan-1-ol acetate	148	21
Pentan-2-ol acetate	134	19
Pentan-3-ol acetate	132	8
2-Methylbutyl acetate	142	28
3-Methylbutyl acetate	142	14

The ester-content of the mixture is over 85%. The alcohols are

obtained by hydrolysis. Pentasol conforms with the following specification: sp. gr. at $20°$ C, 0.812–0.820. b. r. $112°$–$340°$ C.

The following tables give the physical properties of six of the eight
possible amyl alcohols [21].

Amyl alcohol	sp. gr. $(20°)$	b. p. $(°C)$	flash p. $(°C)$	lat. ht.	Binary mixture	
					b. p. $(°C)$	% water
Pentan-1-ol	0.815	137.5	57	102	95.5	47.8
Pentan-2-ol	0.810	119	49.5	98	92.3	32.2
Pentan-3-ol	0.815	115.7	51	97	91.4	32.2
2-Methylbutan-1-ol	0.816	128	63	100	93.8	41.5
2-Methylbutan-2-ol	0.812	101.8	23	93	87.2	22.0
3-Methylbutan-3-ol	0.812	130.5	55.5	100	95.0	42.4

Pentan-1-ol (n-Amyl alcohol) has sp. gr. 0.8296_0, 0.8146_{20}, 0.8110_{25};
m. p. $-79°$ C; $n_D 1.4103$; sp. ht. 0.712; cub. expn. 0.001; visc. 3.68,
3.19_{25}; s. ten. 25.6, 25.15_{25}; dielec. const. 5.1.

3-Methylbutan-1-ol (Isoamyl alcohol) has sp. gr. 0.802835; m. p.
$-117°$ C; $n_D 1.4094$; sp. ht. 0.686; cub. expn. 0.0009; visc. 3.8_{25};
dielec. const. 15.

2-Methylbutan-1-ol (Active amyl alcohol) has sp. gr. 0.8074_{35}; n_D
1.4108; cub. expn. 0.00078; visc. 5.1; opt. rot. $-9.55°$ C.

Pentan-3-ol (see amyl alcohol) has $n_D 1.41$; cub. expn. 0.0015; visc.
4.1_{25}.

2-Methylbutan-2-ol (tertiary amyl alcohol) has $n_D 1.405$; sp. ht. 0.75;
cub. expn. 0.0013; visc. 3.7_{25}.

Solubility of the Amyl Alcohols in Water

	Millilitres of alcohol dissolved in 100 ml water			
	$10°$	$30°$	$50°$	$70°$
Pentan-1-ol	6.4	7.2	8.5	10.7
Pentan-2-ol	8.0	8.8	9.9	11.4
Pentan-3-ol	8.2	9.1	10.2	11.8
2-Methylbutan-1-ol	7.0	7.8	9.2	11.3
2-Methylbutan-2-ol	17.6	17.7	17.8	17.9
3-Methylbutan-1-ol	6.5	7.4	8.7	10.8

Solubility of Water in the Amyl Alcohols

| | Millilitres of water dissolved in 100 ml alcohol | | | |
	$10°$	$30°$	$50°$	$70°$
Pentan-1-ol	2.6	2.1	1.9	1.8
Pentan-2-ol	7.5	5.3	4.4	4.1
Pentan-3-ol	8.0	5.5	4.5	4.2
2-Methylbutan-1-ol	5.0	3.6	3.1	3.0
2-Methylbutan-2-ol	20.5	14.0	10.6	8.7
3-Methylbutan-1-ol	3.7	2.8	2.5	2.4

Synthetic amyl alcohol is not a solvent for cellulose esters or coumarone, but it dissolves ester gum, elemi, mastic, sandarac, kauri, shellac, and is miscible with castor and linseed oils and with hydro-carbons; it partly dissolves dammar.

Fermentation or fusel oil amyl alcohol is available in several grades.
Good commercial quality answers to the following specification: sp. gr. 0.815–0.817; b. r. 128–132° C; flash p. 40–46° C; ign. p. 47° C; m. p. −134; v. p. 9–10; n_D 1.39–1.40; lat. ht. 100; sp. ht. 0.57; cub. expn. 0.00093; visc. 3.3; therm. cond. 0.00037 at 20° C; acidity 0.01% max. as acetic; clearly miscible with benzene in all proportions; water dissolves 3% at 20° C; dissolves 9.6% of water at 20° C.

A less purely technical quality, known as *anhydrous fusel oil*, has the following approximate characteristics: sp. gr. 0.813–0.817; b. r. 105–132° C; flash p. about 36° C. Clearly miscible with benzene in all proportions. This quality contains small quantities of n-propyl, isobutyl and n-hexyl alcohols with traces of pyridine, furfural and esters.

ASTM Specification, D319–58 (1965) for Amyl Alcohol requires: sp. gr. 0.812–0.820 at 20° C; b. r. none below 118° C, max. 5% below 120.0° C, max. 50% below 125.0° C, max. 85% below 130.0° C, none above 140.0° C; max. residue 0.005 g/100 ml; max. H_2O 0.3% by weight; max. acidity 0.005% by weight.

Fermentation amyl alcohol is a solvent for soft copals, ester gum, sandarac, shellac, benzyl abietate, cyclohexanone-formaldehyde and urea-formaldehyde resins. It is miscible with castor and linseed oils and with hydrocarbons. It is a moderately good solvent for ethyl cellulose, fluid silicones, copal ester, mastic and coumarone, and has the property of rendering glyceryl phthalate resins soluble in other solvents. It does not dissolve cellulose esters.

Isoamyl alcohol forms the following binary constant-boiling mixtures:

Isoamyl alcohol (%)	(%)	b. p. ($^\circ$ C)
97.5	Isoamyl acetate 2.5	131.3
26.0	Isoamyl formate 74	123.6
60.0	o-Xylene 40	128.0
53.0	m-Xylene 47	127.0
51.0	p-Xylene 49	126.8
49.0	Ethyl benzene 51	125.9
19	Tetrachlorethylene 81	116.0
36	Chlorbenzene 64	124.3
19	Epichlorhydrine 81	115.4

The following ternary mixtures are known: Isoamyl alcohol 19.6%, isoamyl formate 48%, water 32.4%; b. p. 89.8° C. Isoamyl alcohol 31.2%, isoamyl acetate, 24%, water 44.8%; b. p. 93.6° C.

Another form of synthetic amyl alcohol consists of the secondary alcohols obtained by hydrogenating the higher ketones of ketone or acetone oils (q.v.) these higher ketones, on hydrogenation, give rise to scondary alcohols, the acetates of which have solvent properties similar to the primary alcoholic acetates, butyl and amyl acetates. Hexan-2-ol and 2-methylpentan-4-ol have been produced from olefines. The boiling-points of those of greatest importance are:

	Alcohol ($^\circ$ C)	Acetate ($^\circ$ C)
s-Butyl alcohol	99.5	112
Pentan-2-ol	119	134
Pentan-3-ol	115.7	132
Hexan-2-ol	139	153
Hexan-3-ol	134	150
4-Methylpentan-2-ol	130	148

Secondary butyl alcohol is dealt with separately.

4-*Methylpentan-2-ol* is a hexyl alcohol known industrially as methylisobutyl carbinol. It has sp. gr. $0.807-0.809_{20}$ (pure, 0.808_{20}); b. p. 131.63° C; b. r. 130–133° C; flash p. 115° F (46° C); v. p. 0.9_{10}, 2.1_{20}, 4.06_{30}; n_D 1.4113; lat. ht. 104; s. ten. 22.8; water dissolves 1.7%; dissolves 5.8% of water.

It is a solvent for ethyl cellulose, colophony, ester gum, dammar, mastic, elemi, castor and other vegetable oils.

It has also been proposed to manufacture amyl alcohol and its homologues from the olefines resulting from the cracking of petroleum by absorption in sulphuric acid and hydrolysis of the acid sulphate thus formed [23]. Thus olefine gases are first passed through 75 to 85% sulphuric acid, in order to absorb butylene and amylene, and then through 100% acid to absorb propylene and ethylene, a series of towers being employed. The solutions, on steam distillation, yield the corresponding alcohols (cf. Isopropyl alcohol). The acetates, or other esters, may be obtained directly by treating the alkyl sulphate solutions with calcium or sodium acetate, or with the free fatty acid [24]. These methods are not of industrial importance as regards amyl alcohol, although isopropyl alcohol and secondary butyl alcohol are produced in large quantities by this means (q.v.) the acetate, Methyl anyl acetate is used as a high boiling solvent in cellulose nitrate lacquer systems.

Hexanol. $CH_3 . CH_2 . CH_2 . CH_2 . CH_2 . CH_2 . OH$

Hexan-1-ol (n-hexyl alcohol) is available commercially; it has sp. gr. $0.819–0.821_{20}$; b. p. $157.6°$ C; flash p. $165°$ F ($74°$ C); m. p. $-51.6°$ C; v. p. 0.7; $n_D 1.418$; visc. 4.3_{25}.

The pure alcohol has sp. gr. 0.8198_{20}, 0.8164_{25}; b. p. $156.6°$ C; n_D 1.4174; visc. 5.32, 4.52_{25}; s. ten. 26.55, 26.05_{25}; dissolves 7.2% of water at $20°$ C; water dissolves 0.58% at $20°$ C.

Hexanol is a solvent for shellac, rosin, some gums and dyestuffs, and is miscible with hydrocarbons and vegetable oils. It is without action on rubber.

The following azeotropes are known:

Hexan-1-ol (%)		(%)	b.p. (°C)
18	Ethyl lactate	82	153.6
79	Lemonene	21	155.5
40	Pinene	60	150.8
18	o-Xylene	82	142.3
15	m-Xylene	85	138.3
13	p-Xylene	87	137.7
23	Styrene	77	144
6	Cyclohexane	94	155.7

2-Ethylbutanol $CH_3 . CH_2 . CH(C_2H_5) . CH_2OH$

Characteristics sp. gr. 0.833_{20}; b. p. $149°$ C; flash p. $137°$ F ($58°$ C); n_D 1.43; lat. ht. 109; sp. ht. 0.586; cub. expn. 0.0009; visc. 5.6; s. ten. 28_{28}; water dissolves 0.43%; dissolves 4.6% of water; forms azeotrope with 51% water, b. p. $96.7°$ C.

Heptan-1-ol CH_3 . $(CH_2)_6$. OH

Heptan-1-ol, known also as n-Heptyl alcohol and oenanthyl alcohol has sp. gr. 0.8236_{20}, 0.8202_{25}; b. p.175.9° C; n 1.4238; visc. 7.00, 5.87_{25}; s. ten. 27.25, 26.75_{25}.

2-Ethylhexan-1-ol CH_3 . $(CH_2)_3$. $CH(C_2H_5)$. CH_2 . OH

This octanol known also as Ethyl Hexyl Alcohol is used mainly for the preparation of plasticisers, in particular its phthalates, adipates and sebacates. It is a mild odoured, colourless, slightly viscous, non-toxic liquid and is a solvent for fluid silicones. It is a product of the Fischer-Tropsch process along with lower normal and iso-primary alcohols.

Pure 2-Ethylhexan-1-ol has sp. gr. 0.8344_{20}; n_D 1.1430; lat. ht. 93° C; sp. ht. 0.564; visc. 7; s. ten. 30.

British Standard Specification, BS 1835:1970 for Ethylhexyl alcohol requires: sp. gr. 0.836–0.841 $(0.832–0.837)_{20}$; b. r. 182°–186° C, 95% min; ash. 0.01% max; acidity 0.02% max. as acetic; aldehydes 0.5% max. as ethyl hexaldehyde.

3,5,5-Trimethylhexan-1-ol
$(CH_3)_3$. C . CH_2 . $CH(CH_3)$. CH_2 . CH_2 . OH

This nonyl alcohol is produced with other nonyl alcohols and some octyl and cyclic alcohols by the so-called OXO process in which olefines, such as di-isobutylene, are caused to react with carbon monoxide and hydrogen under high pressure in the presence of a cobalt catalyst to yield aldehydes which hydrogenate to the corresponding alcohols.

3,5,5-Trimethylhexan-1-ol is a colourless, slightly viscous, non-toxic liquid and is a latent solvent for cellulose nitrate, some of the vinyl resins and shellac, it has s. g. 0.8280 at 20/4° C; b. p. about 190–195° C; flash p. 198° F (76° C); n_D1.433; lat. ht. 72.5; cub. exp. 0.00085; visc. 14.1, 3.4_{60}; s. ten. 25.5; water dissolves 0.1%; dissolves 2.9% of water. It forms a pseudo-azeotrope with 83% of water; b. p. 99.5° C.

The phthalate ester (DNP) is used extensively as a primary plasticiser for PVC.

Benzyl Alcohol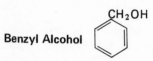

Benzyl alcohol is a widely-used high-boiling solvent which is valued on account of its plasticising effect particularly for cellulose-acetate although this is not permanent. It is particularly useful as a medium in which to grind pigments intended for incorporation in a lacquer, and also for retarding the rate of evaporation of the solvent mixture; it tends to prevent the settlement of pigments. It is a neutral, stable, colourless and almost odourless liquid.

Benzyl alcohol is soluble in water to the extent of 3 to 4% and dissolves about 8% of water at 20° C. n_D 1.538–1.541; b. p. (pure) 205° C; m. p. −15° C; flash p. 205° F (96° C); lat. ht. 111; sp. ht. 0.54; dielec. const. 13; cub. exp. 0.00073; visc. 5.6. It is a solvent for cellulose acetate, for ethyl and benzyl cellulose, zein, ester gum, copal ester, glyceryl phthalate resin, cumarone, benzyl abietate, mastic, shellac, linoxyn, polystyrol, and butadiene-acrylonitrile resins. It is miscible with linseed and castor oils, and with aromatic hydrocarbons, but not with the paraffins or water. Benzyl alcohol has a limited solvent action on cellulose nitrate, nylon and sulphur.

British Standard Specification BS 4192:1967 for benzyl alcohol requires sp. gr. 1.048–1.053, 1.043–1.048$_{25 25}$; b. p. 95% 203–208° C; acidity max. 0.1%; ash max. 0.1%; max. 0.05% combined chlorine; alcohol content not less than 97.0%

Diacetone Alcohol. CH_3\
                          CH_3/ C(OH).CH$_2$.CO.CH$_3$

Diacetone alcohol, diacetone, or 4-hydroxy-4-methylpentan-2-one, is a ketol or ketonic alcohol. It is manufactured from acetone by condensation by means of alkalis and, when pure, is an odourless, colourless liquid. Technical qualities may contain small quantities of acetone and mesityl oxide, a dehydration product; it is to this last impurity that any offensive odour is due.

British Standard Specification 549:1964 [11] for diacetone alcohol requires: sp. gr. 0.940–0.946 (0.936–0.942)$_{20}$; flash p. 40° C. min;

b. r. up to 160° C, 5% max., 160–170° C, 92% min; acidity 0.025% max., as acetic; miscible with 19 volumes of water.

It should be noted that diacetone alcohol slowly decomposes on heating, especially in the presence of even minute traces of alkali [35], and also on long storage, producing acetone; for this reason the figures given above for the boiling range and flash point may not be obtained; the purest material obtainable has the following characteristics: ap. gr. 0.942; b. p. 167° C; n_D 1.424; flash p. 153° F (67° C); m. p. −45° C; v. p. 4; h. c. 8600; visc. 3.6; evap. period 147; cub. exp. 0.00099.

Diacetone alcohol forms an azeotrope with 75% of water b. r. 99.7° C.

Diacetone alcohol is a good solvent for both cellulose nitrate and acetate, also for colophony, bitumen, tars, bakelite, zein, glyceryl phthalate resin, polyvinyl acetate, shellac, benzyl-abietate, kauri, manila, mastic, sandarac, copals and castor oil. It has moderate solvent properties for cellulose ethers, ester gum, polyvinyl chloro-acetate, cumarone, linseed oil, zanzibar, elemi, guaiac. It will not dissolve polyvinyl chloride, rubber or copal ester, and is not compatible with high boiling paraffin hydrocarbons.

Diacetone alcohol is of particular value for cellulose acetate lacquers, since, although a 'high boiler,' it does not remain in the film on evaporation for an undue length of time, as is the case with such high-boiling solvents as cyclohexanol and benzyl alcohol. It gives hard films, having a brilliant gloss, and it is particularly suitable for brushing lacquers destined for use in interiors where odour is objectionable; it is itself non-toxic.

Diethyl Ether. $C_2H_5 . O . C_2H_5$

Ether, known also as Ethoxy ethane [37], is not of itself a solvent for cellulose nitrate, but in admixture with alcohol it has found very extensive application in the manufacture of celluloid. The addition of quite a small proportion of alcohol imparts very powerful solvent properties for cellulose nitrate of high nitrogen content also for polyvinyl acetate. Ether is not a solvent for cellulose acetate, propionate, cellulose ethers, polyvinyl chloride, chloracetate or poly-acrylonitriles.

Ether is produced industrially by the well-known method of treating alcohol with sulphuric acid; hence the name sometimes given to it—sulphuric ether— and from ethylene obtained during the distillation

of petroleum; also by passing alcohol vapour over activated hydrosili-
cates at high temperature [26].

Ether is available in several qualities, the variations being largely due
to the presence of alcohol, water and of the methylating agent in the
alcohol used, other impurities which occur are acetaldehyde, vinyl
alcohol, acetone, acetic acid and the explosive peroxide.

Characteristics (Pure): sp. gr. 0.7199; b. p. 34.6° C; m. p. -117° C;
flash p. -40° F (-41° C). Auto-ignition temp. 186° C; n_D 1.356; sp. ht.
0.538; lat. ht. 84; ht. comb. 8807; s. ten. 17; v. p. 440; cub. exp.
0.00164; visc. 0.234; elec. cond. 3.7×10^{-13}; dielec. const. 4.4.
Dissolves 1.3% of water at 20° C. Water dissolves 6.9% at 20° C.

Ether forms a constant-boiling mixture with ethyl alcohol, boiling at
74.8° C and containing 40% of ether.

British Standard Specification, BS 579:1957 for technical ether
requires: sp. gr. 0.719–0.725, 0.714–0.720$_{20}$; b. r. 95.0% min. below
36.0° C; max. residue 0.005%; max. acidity 0.002% as sulphuric;
peroxides, absent. (See appendix D of BS 579:1957 for potassium
iodide test).

Ether is a very dangerous substance to handle. It is highly narcotic,
and large doses readily lead to death; it forms explosive peroxides.
Ether can be protected from peroxidation by storage in the presence of
iron, a large measure of protection is given by storage in the dark [34]
and in the presence of 20% sodium hydrate solution. It is exceedingly
inflammable. Mixtures with air can explode with great violence. The
explosive limits (see p. 56) lie between 35 g and 200 g per cubic metre
of air with a maximum explosivity at about 120 g. Ether very readily
acquires high static charges of electricity which may give rise to sparks
resulting in the ignition of the ether; for this reason it is advisable to
'earth' the contents of glass vessels from which any large quantity of
ether is to be poured.

Di-isopropyl ether. $C_3H_7 . O . C_3H_7$

Di-isopropyl ether is used industrially for extractions at low temperatures. It is an exceedingly dangerous substance as it forms unstable
peroxides, on contact with air, which explode with great violence when
heated [29]; this tendency can be limited by storing in the dark and by
adding small proportions of stabilisers such as catechol, hydroquinone,
aqueous sulphites, sodium amalgam, or other reducing agents. It can be

maintained almost free from peroxide by storage over 20% sodium hydrate solution [36].

Di-isopropyl ether is a solvent for hydrocarbons, paraffin wax, vegetable, animal, essential and mineral oils, colophony, ester gum and rubber, but not for benzyl cellulose, cellulose acetate, shellac, kauri, or polyvinyl resins. It is a moderate solvent for dammar and ethyl cellulose and for cellulose nitrate when mixed with anhydrous alcohols.

It is used in place of ethyl ether for extracting alkaloids, vitamins, nicotine and for dewaxing lubricating oils, for propellants and for high octane aviation fuels.

Characteristics: sp. gr. 0.730, 0.7247_{20}; b. p. 68° C; n_D 1.368; flash p. -7° F (-22° C); v. p. 119; m. p. −85° C; sp. ht. 0.526; lat. ht. 68; s. ten. 32; visc. 0.34_{20}, 0.38_{25}; coeff. exp.0.00145; water dissolves 0.65%; dissolves 0.9% of water. It forms azeotropes with 4.5% of water (b. p. 62.2° C) and with 14% of isopropanol (b. p. 66.2° C). It is narcotic but not toxic.

Di-n-butyl ether. $C_4H_9 . O . C_4H_9$

This is a solvent for fats, oils, ester gum, rosin, dammar, kauri, rubber and hydrocarbons, but not for cellulose esters or benzyl cellulose. It is a useful alternative to diethyl ether.

sp. gr. 0.770 at 20° C; b. p. 142° C; m. p. −95° C; v. p. 5; flash p. 87° F (31° C); lat. ht. 68; n_D 1.399; s. ten. 23; visc. 0.32; water dissolves 0.03%; dissolves 0.19% of water.

It forms explosive peroxides [27] similar to those mentioned above and has a mild odour.

Di-methyl Acetal. $CH_3 .CH(OCH_3)_2$

A product of about 70% purity is available industrially [32] it has sp. gr. 0.845–0.847; b. r. 57–60° C (95%) and contains about 30% of methanol and 2% of acetaldehyde. Pure dimethyl acetal has b. p. 63.6; n_D 1.366; sp. gr. 0.852_{20}. It dissolves 4.3% of water and water dissolves 28% at 20° C. It forms an azeotrope with 25% of methanol (b. p. 56.8° C) and another with 5% of water (b. p. 59° C). It is miscible with all the usual solvents and dissolves ethyl cellulose, ester gum, dammar, coumarone, polystyrene but not cellulose esters, shellac or polyvinyl resins.

It is highly inflammable and rapidly produces narcosis in high concentration.

Acetal. $CH_3 . CH . (OC_2H_5)_2$

Acetal, or acetaldehyde diethyl acetal, does not dissolve cellulose acetate and is a poor solvent for celluose nitrate, admixture with anhydrous alcohol greatly increases its solvent power.

Characteristics: b. p. $102°$ C; sp. gr. 0.826_{20}; lat. ht. 66; sp. ht. 0.52; n 1.38; dielec. const. 2.8 at $25°$ C; water dissolves about 5%.

Acetal forms a constant-boiling mixture with 67% of ethyl alcohol boiling at $79°$ C.

Acetal solvent [28] consists of a mixture of acetal and alcohol which dissolves cellulose nitrate, kauri, manila and other resins. It has b. r. $75°-85°$ C (90%); sp. gr. 0.82; flash p. $38°$ F ($3°$ C); acidity 0.01% max.

Paraldehyde.

Paraldehyde is a medium boiling solvent [29] for ethyl cellulose, polystyrene, colophony and coumarone, it partly dissolves ester gum, dewaxed dammar and copals and is a latent solvent for cellulose nitrate; its solvent power for the latter is activated by the addition of non-solvents such as alcohol and by solvents such as acetone; its precipitating effect and evaporation rate are about one half that of toluene. It does not dissolve cellulose acetate or acetobutyrate, polyvinyl chloride or copolymers.

Paraldehyde is a powerful narcotic but is believed not to be definitely toxic, it possesses a somewhat choking odour. It slowly decomposes to acetaldehyde, and its flashpoint lowers and its acetic acid content rises in consequence; stabilizers are sometimes added to prevent this decomposition.

It has sp. gr. 0.998; n_D1.405; flash p. $81°$ F ($27°$ C); m. p. $12°$ C); visc. 1.2.

The British Pharmacopoeia requires sp. gr. 0.998–1.0; m. p. $11°$ C min.

References

[1] F.P. 354,621.
[2] *Cf.* Hilditch, "Catalytic Processes in Applied Chemistry." Chapman and Hall Ltd.
[3] G.P. 293,787.

[4] *Compt. rend.*, 1924, **179**, 1330.
[5] *Chim. et. Ind.*, 1925, p. 179.
[6] F.P. 593,650.
[7] F.P. 594,121; 593,649.
[8] *Ind. Eng. Chem.*, 1925, p. 43.
[9] Brit. Pat. 245,991.
[10] G.P. 367,204.
[11] *Cf.* Brit. Pat. 7726 (1910); 109,993 (1916); 162,370. *Chem. and Ind.*, 1927, p. 456T.
[12] *Chem. and Ind. Review*, May 28th, 1926.
[13] Brit. Pat. 146,957; 249,834; 248,395; 303,176. U.S. Pat. 1,365,056; 1,365,050; 1,879,660. *Ind. Eng. Chem.*, 1926, p. 844. *Chem. Met. Eng.*, 1926, p. 400.
[14] *Cf.*490; 349,491; 409,730; 415,311; 415,312; 428,523; 437,120; 437,121; 450,530; also *Ind. Eng. Chem.*, 1947, p. 1445.
[15] *Chim. et. Ind.*, 1925, p. 179.
[16] F.P. 594,121; 593,649.
[17] Brit. Pat. 282,448; 326,812; 336,811; 364,134.
[18] The Sharples Solvents Corporation, Philadelphia, U.S.A. *Cf.* Aschan, *Chem. Ztg.*, 1918, 11, 939,955; Clark. *Ind. Eng. Chem.*, 1930, p. 439. Gillette and Price. *Ind. Eng. Chem.*, 1950, p. 2388.
[19] Kock and Burrell. *Ind. Eng. Chem.*, 1927, p. 442; Kirkpatrik. *Chem. Met. Eng.*, 1927, p. 176; *Chem. Trade J.*, 1927, p. 417.
[20] Sharples Chemicals Inc., Airco Export Co., N.Y., U.S.A.
[21] Ayres, T. *Am. Inst. Chem. Eng.*, 1929, **22** 23.
[22] Carbide and Carbon Chemicals Co., U.S.A. and Shell Chemicals, London, W.C.2.
[23] U.S. Pat. 1,365,046; Brit. Pat. 248,395; 249,834.
[24] U.S. Pat. 1,365,050; 1,365,052.
[25] U.S. Pat. 1,823,704.
[26] U.S. Pat. 1,908,190.
[27] *Chem. and Ind.*, 1933, p. 274; 1936, p. 241, 580. *J. Chem. Educ.*, 1940, , 595. *Brit. Chem. Abs.*, 1939, A11, p. 531. *Ind. Eng. Chem.*, 1940, p. 124. *Chem. Trade J.* 1943 p. 67. *J. Soc. Chem. Ind.*, 1946, p. 421.
[28] Howards & Sons Ltd., London, e. *Cf.* B.P. 357,277; U.S. Pat. 1,860,822.
[29] Brit. Pat. 22,540 (1896); 367,390 (1930); U.S. Pat. 996,191 (1911); German pat. 195,312 (1906); 343,162 (1921); 364,347 (1922).
[30] B. D. Sully. Private communication.
[31] H.M. Stationery Office, London, W.C.2.
[32] British Industrial Solvents Ltd., London, S.W.1.
[33] E. J. Lush. Brit. Pat. 203,218 (1922).
[34] W. Hunter and J. Downing. *J. Soc. Chem. Ind.*, 1949, **68**, 362.
[35] R. P. Bell and J. E. Prue. *J. Chem. Soc.*, 1949, p. 362.
[36] Imperial Chemical Industries Ltd., London, S.W.1.
[37] Shell Chemicals Ltd., London, W.C.2.
[38] International Union of Chemistry.

CHAPTER THREE

Ketones

Acetone

$$\begin{matrix} CH_3 \\ \quad\quad C{=}O \\ CH_3 \end{matrix}$$

Acetone or dimethyl ketone or propan-2-one is one of the most widely used of the industrial solvents on account of its low cost, high solvent powers and comparative lack of toxicity. Its principal uses are as a low boiling solvent for cellulose acetate and nitrate lacquers, as a medium for preparing acetate silk, for artificial leather manufacture, for dewaxing lubricating oils, for the manufacture of cordite explosive, photographic film, safety glass, and for dissolving acetylene.

The various routes by which acetone can be manufactured are briefly as follows:

(1) Up to about 1914 the chief method was that of the distillation of acetate of lime [1, 2] obtained by the destructive distillation of wood. This method is now obsolete.

(2) Isolation from crude wood spirit, also an obsolete method.

(3) The *Clostridium* fermentation of carbonhydrates producing also butyl alcohol [3, 4].

(4) Together with higher ketones by distilling the calcium salts of the acids obtained by fermenting the giant seaweed which grows along the North Pacific Coast [5]. (obsolete)

(5) The hydration of acetylene to acetaldehyde oxidation to acetic acid and pyrolysis over alkaline earths [6]. A method of great importance during the 1914–18 war but now discontinued.

(6) The interaction of acetylene and water at high temperature in the presence of zinc oxide catylist.

(7) Catalytic interaction of alcohol and steam [16] in the presence of iron-copper-manganese catalystat 470° C. A method used very successfully between the wars but discontinued in 1946.

(8) As a co-product during the production of phenol from cumene (isopropyl benzene) a method recently introduced [14].

(9) Dehydrogenation of isopropanol is the largest commercial process in modern operation since the isopropanol itself is readily obtained from hydrated propylene available in commercially cheap quantities from cracked petroleum [20].

136

Characteristics: The purest acetone has the following properties: sp. gr. 0.7899$_{20}$; n_D 1.3599; visc. 0.32; b. p. 56.3; m. p. −94.9; v. p. at 20° C 186 mm; flash p. 2° F (−16.7° C); lat. ht. 122; sp. ht. 0.528; ht. comb. 7373; cub. exp. 0.00149; s. ten. 24; elec. cond. 1×10^{-7}; dielec. const. 21; dil. rat., benzene 5.2, toluene 4.5, butanol 7.0.

British Standard Specification, BS 509:1964, for acetone requires: sp. gr. 0.796−0.798 (0.791−0.793$_{20}$. b. r. 55°−56° C; Acidity 20 ppm max. as acetic acid; alkalinity to methyl red−*nil*; residue 30 ppm max. Withstands permanganate test for two hours minimum, and alcoholic impurities test for five minutes minimum; max. H$_2$0 0.50% by weight.

Cordite acetone must answer to the following specification: sp. gr. at 15° C not over 0.800. Free from colour, clearly miscible in all proportions with water. Acidity 0.002% max. as CO$_2$. Fixed acidity, nil. Aldehydes or other reducing substances, 1 part per thousand maximum. Must withstand the permanganate test for at least thirty minutes (viz., 1 cc of a 0.1% solution of KMnO$_4$ added to 100 cc of acetone kept at 15° C in the dark, the colour to persist for at least thirty minutes).

ASTM Specification, D 329−66: (99.5% grade), sp. gr. 0.7910−0.7930 at 20° C; b. r. within 1 deg C including 56.1° C; max. residue 0.005g/100ml; max. H$_2$O 0.5% by weight; max. acidity 0.002% by weight; max. alkalinity −0.001% by weight; miscible with distilled H$_2$O in all proportions; to withstand permanganate test for 30 min.

Acetone is one of the most powerful solvents known. It dissolves cellulose acetate, tripropionate, benzoate, nitrate, aceto-nitrate, aceto-propionate, aceto-butyrate, ethyl cellulose, glyceryl phthalate resins, thiourea resin, croton-aldehyde resin, cyclohexanone-formaldehyde resin, benzyl abietate, polyvinyl acetate and chloro-acetate, low molecular weight polystyrene, low viscosity silicones, cumarone, ester gum, copal ester, and rubber chloride. It is miscible with castor, linseed, china-wood, cotton seed, fish, olive, palm, peanut, neatsfoot oil, tallow, and polymerised oils, and with most hydrocarbons. It attacks benzyl cellulose, polythene, polyvinyl chloride, methyl methacrylate. Its solvent action on rubber varies widely with the rubber [8], Balata being the most insoluble. It does not dissolve nitro-polystyrene, polyacrylo-nitrile, nylon.

The solubilities of natural resins, waxes and pitches are approximately as follows:

Solubility 90% *or more.* Beeswax, candelilla wax, elemi, dammar, Japan wax, soft manila, mastic, pontianac, rosin, sandarac, shellac.

Solubility 50–90%. Copal, kauri, montan wax, ozokerite refined coal-tar pitch, asphalt.

Solubility 20–50%. Carnauba wax, Congo copal, gilsonite, hard manila, refined Trinidad pitch, Zanzibar copal, gutta-percha, Balata.

Solubility below 20%. Arabic, tragacanth, Syrian asphalt.

Acetone forms the following azeotrope mixtures:

Acetone	%	b. p. (°C)
86.5	Methyl alcohol 13.5	55.95
34	Carbon disulphide 66	29.25
20	Chloroform 80	64.7
77	Carbon tetrachloride 23	55.8
55	Methyl acetate 45	56.1

the last mixture has a boiling-point higher than that of either of its components.

Sp. gr. at 20° C of acetone-water mixtures [10]:

Acetone (%)	sp. gr.	Acetone (%)	sp. gr.
10	0.986	60	0.900
20	0.974	70	0.877
30	0.958	80	0.852
40	0.942	90	0.825
50	0.921	100	0.795

'Methyl Acetone' and 'Wood Spirit'

In the manufacture of acetone and methyl alcohol from wood, fractions are obtained containing methyl acetate and other substances which cannot economically be separated one from another; these fractions have excellent properties and are recognized articles of commerce, but they vary somewhat widely in composition and are not largely used. The composition of 'Methyl acetone' varies usually between the following limits: methyl acetate, 20–30%; acetone, 35–60%; methyl alcohol, 20–40%. A good commercial quality [12] has sp. gr. 0.833–0.838; b. r. 95% 54–70° C; flash p. 10° F (−8° C); acidity 0.02% max; Clearly miscible with water and with carbon disulphide.

Wood spirit, known also as wood naphtha and as wood alcohol, contains about 80% of methyl alcohol and varying quantities of acetone, methyl acetate, acetal, allyl alcohol, acetaldehyde and methylamine.

Methyl acetone is an excellent solvent for cellulose acetate and nitrate, for ester gum, benzyl abietate, and is miscible with castor and linseed oils, hydrocarbons, and water.

Wood spirit is an exceedingly rapid solvent for cellulose nitrate; it also dissolves cellulose acetate, ester gum, benzyl abietate, coumarone, castor and linseed oils. It partially dissolves shellac, mastic, copal, and is miscible with hydrocarbons and with water.

Methylethyl Ketone. $CH_3.CO.C_2H_5$

Methylethyl ketone or butan-2-one is an important commercial solvent, extractant, and chemical intermediate, (e.g. antioxidants, perfumes, catalysts). It is mainly obtained from petroleum sources (e.g. dehydrogenation of secondary butyl alcohol) and lesser amounts are prepared by the destructive distillation of wood.

The formation of higher ketones from lower ketones and alcohols has been accomplished by passing the appropriate mixture with hydrogen over aluminium oxide catalyst at temperatures ranging from $150°-400°$ C; thus acetone and methyl alcohol at $230°$ C yield methylethyl ketone, diethyl ketone and higher ketones as the main products.

British Standard Specification, BS 1940:1968 for methylethyl ketone requires: sp. gr. 0.809 − 0.811 (0.805 − 0.807 at $20°$ C); b. r. not less than 95% between $79°$ C and $80.5°$ C; max. residue 0.005% by weight; no opalescence with 19 volumes of carbon disulphide at $15°$ C; max. acidity 0.003% by weight; max. alcoholic impurities 1.0% by weight.

ASTM Specifications, D 740−66 for methylethyl ketone requires: sp. gr. 0.805 − 0.807 at $20°$ C; b. r. $78.5°$ − $81°$ C; max. residue 0.005 g/100 ml; max. H_2O 0.3% by weight; max. acidity 0.005% by weight; alcohol (as s-butyl alcohol) not more than 0.7% by weight.

This quality is a good solvent for cellulose nitrate, cellulose aceto-butyrate and for ethyl cellulose, fluid silicones, kauri, manila, pontianac, colophony, ester gum, benzyl abietate, plextol, glyceryl phthalate resin, cumarone, dewaxed dammar, rubber chloride. Its solvent power for cellulose acetate varies with variations in the latter, but in general MEK is not a good solvent for cellulose acetate. It does not dissolve polyacrylonitrile, carnauba, beeswax, ceresine or paraffin wax. It is miscible with castor and linseed oils and with hydrocarbons.

Methylethyl ketone finds use in adhesives based on 'Neoprene' and butadiene/acrylonitrile rubbers as well as those based on cellulose nitrate, vinyl copolymers, polystyrene, and acrylic resins.

The following azeotrope mixtures are known [9]:

Methylethyl ketone %	(%)	b. p. (°C)
88.6	Water 11.4	73.5
15	Carbon disulphide 85	45.8
29	Carbon tetrachloride 71	73.8
45	Benzene 55	78.5
69	Methyl alcohol 31	64.1
66	Ethyl alcohol 34	74.8
68	Isopropyl alcohol 32	77.5
52	Methyl propionate 48	79.25
55	Propyl formate 45	79.45
22	Ethyl acetate 78	76.7
73	Tertiary butyl alcohol 27	77.5
40	Cyclohexane 60	72.0

Two ternary azeotropes are known, one with 74.8% carbon tetrachloride and 3.0% water (b. p. 65.7° C) and the other with 73.8% benzene and 8.9% water (b. p. 68.9° C).

Pure methylethyl ketone has n1.379; b. p. 79.6°C); m. p. -86°C; flash p. 19°F (-7°C); lat. ht. 104; sp. ht. 0.55; v. p. at 20°C, 77.5; sp. gr. 0.8050_{20},cub. exp. 0.0013; s. ten. 25; dielec. const. 18; elec. cond. 5×10^{-8}; dil. rat. 4.5; Its miscibility with water varies greatly with temperature [13].

$$\text{Ethyl amyl ketone} \quad CH_3.CH_2.\overset{\overset{\textstyle O}{\|}}{C}.CH_2.\overset{\overset{\textstyle CH_3}{|}}{CH}.CH_2.CH_3$$

Ethyl amyl ketone or 5-methylheptan-3-one is a high boiling ketone with an agreeable ketonic odour. Ethyl amyl ketone possesses the following characteristics: sp. gr. 0.822_{20}; b. p. 160.5° C; flash p. 136° F (57.9° C); n_D 1.4149; dil. rat. toluene 2.2.

Ethyl amyl ketone is a high-boiling solvent for nitrocellulose, cellulose aceto-butyrate, ethyl cellulose, resin, ester gum, (glycerol and pentaerythritol) congo ester, coumarone-indene, chlorinated rubber, vinyl chloride/vinyl acetate copolymers; vinyl chloride/vinylidene copolymers, polyvinyl acetate, polystyrene, acrylic, and sulphonamide resins.

Its slow evaporation rate assists in obtaining good secondary flow and its low water solubility has led to its inclusion in two-phase multi-colour finishes. It is also a useful chemical intermediate and its inherent stability under normal conditions of storage and use eliminates any risk of acidic decomposition products. Its toxicity is low but normal precautions should be taken when handling continuously.

Diethyl ketone $C_2H_5.CO.C_2H_5$

Diethyl ketone, known also as amyl ketone, pentan-3-one, propionone and dimethyl acetone. The pure substance has sp. gr. 0.818; b. p. 101° C; cub. exp. 0.00113; visc. 47. It dissolves 2% of water at 20° C. It is produced by the interaction of ethylene and hydrogen with a large excess of carbon monoxide at about 200° C under very high pressures. The commercial product is a colourless liquid having a fruity camphoraceous odour, and consists of a mixture of diethyl ketone and methyl-n-propyl ketone with about 2% of non-ketonic impurities. It has sp. gr. $0.807-0.811_{20}$; b. r. 90% between 100°–105° C; s. ten. 25; visc. 0.47 at 25° C; dielec. const. 15.

Methyl-n-propyl Ketone , $CH_3.CO.C_3H_7$

Methyl-n-propyl ketone, or pentan-2-one, has sp. gr. 0.812; b. p. 101.7° C; n_D 1.3895; v. p. 30 mm. at 20° C; flash p. 45° F (7° C); lat. ht. 91; m. p. −83°C; visc. 0.5; dissolves 3.6% of water at 25° C; water dissolves 6% at 25° C; dil. rat. toluene 4.4; s. ten. 25; dielec. const. 15.

Butyrone $C_3H_7.CO.C_3H_7$

Butyrone, heptan-4-one, or di-n-propyl ketone is manufactured from wood, which is saccharified and then fermented in the presence of chalk, whereby impure calcium butyrate is produced. Calcium butyrate on dry distillation yields butyrone. The wide boiling range of the technical product is accounted for by the fact that aliphatic acids other than butyric are also produced, a variety of ketones consequently arises on distilling the calcium salt.

   Butyrone is a poor solvent for both cellulose acetate and for nitrate; it dissolves caoutchouc, rubber chloride, ester gum, cumarone, linseed and castor oils, glyceryl phthalate and other resins. It is miscible with hydrocarbons. It possesses a rather sweet odour resembling pineapples, and is non-toxic at normal concentrations.

   Technical butyrone has sp. gr. 0.820; b. r. 137°–144° C; flash p. 120° F (49° C); n_D 1.41; visc. 0.74; sp. ht. 0.55; lat. ht. 76; v. p. 5; s. ten. 25 at 25° C; cub. expn. 0.0011; water dissolves 0.53%; dissolves 1.27% of water.

Di-isopropyl ketone $(CH_3)_2.CH.CO.CH.(CH_3)_2$

A colourless liquid with a strong and somewhat unpleasant odour. It is a solvent for cellulose nitrate but not for cellulose acetate, formvar or nylon. It dissolves ethyl cellulose and polyvinyl chloro-acetate on heating. It has sp. gr. 0.808_{20}; b. p. $125.5°$ C; b. r. $120-127°$ C.

Methyl-n-butyl Ketone $CH_3.CO.C_4H_9$

Methyl-n-butyl ketone, or hexan-2-one, has sp. gr. 0.817; b. p. $127°$ C; m. p. $-56°$ C; n_D 1.4024; v. p. 10 mm. at $20°$ C; flash p. $73°$ F ($23°$ C); lat. ht. 83; sp. ht. 0.553; visc. 0.63; dissolves 3.7% of water at $25°$ C; water dissolves 3.5% at $25°$ C; dil. rat. toluene 4.1; cub. expn. 0.001; s. ten. 25; dielec. const. 12.

Methyl isobutyl ketone. $(CH_3)_2.CH.CH_2.CO.CH_3$

Methyl isobutyl ketone or 4-methylpentan-2-one is generally known as MIBK. It is a medium boiling solvent possessing a bland ketonic odour, high solvent power, and an economic price. It is generally obtained from petrochemical sources by the hydrogenation of mesityl oxide [16] (q.v.). Sp. gr. $0.800-0.804_{20\ 20}$; b. p. $116°$ C; m. p. $-85°$ C; n_D 1.396; s. ten. 25; sp. ht. 0.46; lat. ht. 87; dil. rat. toluene 3.6; flash p. $73°$ F ($23°$ C) open; v. p. 15; dissolves 2.8% of water at $20°$ C; water dissolves 1.9% at $20°$ C; visc. 0.58; forms a pseudo-azeotrope with 24.3% of water boiling at $87.9°$ C.

British Standard Specification, BS 1941:1953, for methyl-n-butyl ketone, requires: sp. gr. $0.803-0.807$, $(0.800-0.804_{20})$; b. r. 95% between $114°$ C and $117°$ C; residue 0.01% max.; clear with 19 vols CS_2; acidity 0.01% max. as acetic; alcohols 1% max. as hexanol.

Methyl isobutyl ketone is a solvent for colophony, rubber chloride, rubber, polyvinyl chloride, acetate and chloro-acetate, dammar, ester-gum, elemi, mastic, sandarac, ethyl cellulose, benzyl cellulose, cellulose nitrate; it is miscible with mineral and vegetable oils. It partly dissolves kauri, zanzibar, pontianac, manila, beeswax, paraffin wax and blown oils; it is a non-solvent for shellac, tragacanth, carnauba wax and cellulose acetate. Its chemical stability has attracted interest in one-pack moisture cured polyurethane finishes and this same property encourages its use during the processing of such widely differing products as penicillin and acetic acid. MIBK is also employed in dewaxing mineral oils.

Methyl-n-amyl ketone. $CH_3.CO.C_5H_{11}$

Methyl-n-amyl ketone, or heptan-2-one, has sp. gr. 0.820; b. p. 150° C; n_D 1.410; v. p. 3 mm. at 20° C; flash p. 106° F (41° C); lat. ht. 83; visc. 0.81; dissolves 1.5% of water at 20° C; water dissolves 0.43% at 20° C; dil. rat. toluene 3.9; m. p. −35° C.

This ketone is available technically with the following characteristics: sp. gr. $0.816–0.821_{20}$; b. r. 147°–154° C; acidity 0.05% max. as acetic acid; dil. rat. toluene 3.9.

Its solvent properties are similar to those of methyl isobutyl ketone.

Di-isobutyl Ketone.
$(CH_3)_2.CH.CH_2.CO.CH_2.CH.(CH_3)_2$

Di-isobutyl ketone or 2,6-dimethyl/heptan-4-one is a stable colourless liquid having a faint odour like that of amyl acetate. It has sp. gr. 0.809 at 20° C; b. p. 168° C; flash p. 120° F; n_D 1.41; v. p. 1.7; dil. rat. xylene 1.5; water dissolves 0.08% and it dissolves 0.46% of water at 20° C.

It dissolves cellulose nitrate, colophony, estergum, dammar, chlorinated diphenyls,and some alkyl, methacrylate and phenolic resins but not shellac, cellulose acetate, ethyl cellulose, or methyl methacrylate but its action on polyvinyl acetate, chloride and chloroacetate depends largely on their degree of polymerisation. It is miscible with most of the vegetable oils and usual solvents.

Mesityl Oxide

$$\begin{array}{c} CH_3 \\ \diagdown \\ \diagup \\ CH_3 \end{array} C:CH.CO.CH_3$$

Mesityl oxide or 4-methylpent-3-en-2-one is an unsaturated ketone obtained by dehydrating diacetone alcohol, or prepared direct from acetone by means of anhydrous alkalis, such as lime, or dehydrating agents, such as zinc chloride or hydrochloric acid gas. It has a rather strong odour suggestive of mice and peppermint, which precludes its extensive use. Mesityl oxide occurs in acetone oils along with a further condensation product of acetone, phorone, $(CH_3)_2:C:CH.CO.CH:C:(CH_3)_2$. b. p. 197° C.

Mesityl oxide is a solvent for low viscosity cellulose acetate and nitrate and yields tough, transparent and glossy films. It dissolves cellulose aceto-butyrate, ethyl cellulose, ester gum, colophony, kauri, raw rubber, polyvinyl chloride and acetate and co-polymers, coumarone and most oils; it partly dissolves shellac and dammar. Were it not for its disagreeable odour, it would undoubtedly be used extensively as a solvent for cellulose acetate. It has been recommended as a paint-stripper [12].

Characteristics; The pure substance has sp. gr. 0.857_{20}; b. p. 129° C; m. p. −59° C; v. p. 8.7; flash p. 82° F (27.8° C); sp. ht. 0.52; lat. ht. 86; visc. 0.88 at 25° C; n_D 1.444; cub. exp. 0.0011; dielec. const. 15; water dissolves 3.4%; dissolves 3.4% of water. Good technical material [12] has n_D 1.44−1.45; flash p. 75° F (24° C); b. r. 95% between 120° C and 130° C.

Mesityl oxide forms an azeotrope with 30% of water b. p. 91.5° C.

References

[1] F. P. 439,732 (1911).
[2] *Cf.* Young's "Distillation Principles and Processes," 1922. Macmillan & Co. Ltd.
[3] Brit. Pat. 137,538.
[4] Brit. Pat. 21,073 (1912); 4,845 (1915).
[5] Dyson. *Chem. Age,* p. 390.
[6] Canadian Electro-Products Co., Shawinigan.
[7] U.S. Pat. 1,497,817. *Chem. and Ind.,* 1930, p. 53 T.
[8] *Cf* Remler. *Ind. Eng. Chem.,* 1923, p. 717.
[9] Langedijk. *Chem. and Ind.,* 1938, 891.
[10] Naville. *Helv.* Chim. Acta, 1926, p. 913.
[11] Carbide and Chemicals Co., N.Y., U.S.A.: Stemoc Ltd., London.
[12] British Industrial Solvents Division of The Distillers Co., Ltd., London, W.1.
[13] Marshall. *J.C.S.,* 1906, p. 1381.
[14] Distillers Co., Ltd. and Hercules Powder Co., Inc.
[15] Commercial Solvents (G.B.) Ltd.
[16] Shell Chemicals Ltd., London, W.C.2.

CHAPTER FOUR
Esters

Methyl Acetate. $CH_3 . COO . CH_3$

This ester is similar to acetone in character, the boiling-points of the two liquids being the same and their solvent powers comparable. Methyl acetate is not widely used, since it readily becomes acid on contact with water, and the fact that its parent alcohol—methyl alcohol—is poisonous also militates against its industrial employment. It is highly inflammable. It is chiefly used to replace acetone when expediency demands.

Pure methyl acetate has sp. gr. 0.933, $0.9342_{20\ 4}$, $0.9279_{25\ 4}$; b. p. 56.9° C; $n_D 1.3614$; m. p. $-98°$ C; flash p. 8.6° F ($-13°$ C); sp. ht. 0.50; lat. ht. 0.98; elec. cond. 3.4×10^{-6}; dielec. const. 7.4; cub. exp. 0.00134; s. ten. 24.8, 24.1_{25}; visc. 0.385, 0.364_{25}; ht. comb. 5371.

The following azeotropic mixtures are of interest:

Methyl acetate (%)	(%)	b. p. (°C)
81	Methyl alcohol 19	54.0
30	Carbon disulphide 70	40.2
45	Acetone 55	56.1
30	Isopropyl alcohol 70	77.3

It is a good solvent for cellulose nitrate and acetate, for ester gum, colophony, kauri, manila, sandarac, polyvinyl acetate, benzyl abietate, glyceryl phthalate resin, and is miscible with castor and linseed oils and with hydrocarbons. It is a solvent for ethyl cellulose [1] when mixed with other substances such as camphor, triphenyl phosphate, cyclohexanol, ethyl benzene, acetophenone, butyl tartrate, cyclohexanone, dibenzylamine, dimethyl aniline, ethyl benzyl aniline, diphenylmethane, monochloro-napthalene.

It does not dissolve copal ester gum, shellac, dammar, polyvinyl chloride, elemi or coumarone. It is miscible with water; at 20°C water will dissolve 24.35% by weight of methyl acetate. It is made by esterification, in the liquid phase, or by passing the mixed vapours of methyl alcohol and acetic acid over heated catalysts [2].

Ethyl Acetate. $CH_3 . COO . C_2H_5$

This solvent is undoubtedly the best of the low-boiling solvents for the manufacture of cellulose nitrate lacquer. It has considerable advantages

145

over acetone; its boiling-point is higher and its evaporation rate about one half that of acetone, and it is therefore less likely to cause chilling. Its dilution ratios for cellulose nitrate solutions against dilution with alcohols and paraffins are larger than those of acetone, although smaller against aromatic hydrocarbons. Pure ethyl acetate is not a solvent for cellulose acetate, 5 to 30% of uncombined alcohol being necessary to effect solution, and for this reason ethyl acetate containing up to 15% of dry alcohol is frequently employed for both cellulose acetate and nitrate, the presence of free alcohol also improving its solvent powers for the nitrate. Solutions of cellulose nitrate in ethyl acetate have viscosities about one-third of solutions of corresponding concentration in butyl acetate, and one-fifth of those in amyl acetate; such solutions have tolerances for dilution with butyl alcohol, toluene or xylene approaching those in ethyl lactate.

Ethyl acetate is mainly manufactured by direct continuous esterification; it is also made by the catalytic condensation of acetaldehyde by means of alkoxides and by the interaction of ethylene and acetic acid at $150°-170°$ C in the presence of a boroflouride catalyst.

Ethyl acetate is available commercially in several grades.

British Standard Specification, BS 553:1965, for ethyl acetate requires: sp. gr. 0.905–0.908 (0.900–0.903 at 20° C); esters 99% min; b. r. 95% between 76.5° C and 78.5° C; residue 0.01% max; acidity 0.01% max; Clearly miscible with 19 volumes of carbon disulphide at 15° C.

ASTM Specification, D302–58 (1965) for 85–88% grade requires: sp. gr. 0.882–0.887 at 20° C; b. r. $71.0°-79.0°$ C; max. residue 0.005 g/100 ml; max. H_2O 0.2% by weight, miscible with 19 volumes of 99% heptane at 20° C; max. acidity 0.01% by weight; ester value 85.0–88.0 by weight.

Pure ethyl aetate [14] has the following characteristics: sp. gr. 0.9066, $0.9007_{20\ 4}$, $0.8946_{25\ 4}$; b. p. 77.15° C; v. p. at 20° C 73 mm; flash p. 25° F (−4° C); m. p. −83.6° C; evap. period 2.9; cub. exp. 0.0013; lat. ht. 0.87; sp. ht. 0.48; n_D1.3728; s. ten. 23.95_{20}, 23.3_{25}; visc. 0.452_{20}, 0.425_{25}; dil. rat. benzol 3.4, toluene 3.4, butanols 8.4; elec. cond. 3.2×10^{-7}; dielec. const. 6.11; ht. comb. 6103.

	0°	10°	20°	30°
Water dissolves %	10.1	8.9	7.9	7.2
Dissolves water %	2.3	2.6	3.0	3.5

The following azeotropic mixtures are known:

Ethyl acetate (%)	(%)	b. p. (°C)
69.4	Ethyl alcohol 30.6	71.8
77.0	Isopropyl alcohol 23	74.8
20.0	Methyl alcohol 80	62.3
93.9	Tert. butylalcohol 6.1	70.4
3.0	Carbon disulphide 97	46.1
43.0	Carbon tetrachloride 57	74.7

and a ternary mixture consisting of ethyl acetate 83.2%, ethyl alcohol 9%, water 7.8%, boiling at 70.3° C. It forms a pseudo-azeotrope with 8.2% of water boiling at 70.45° C.

Ethyl acetate is a good solvent for cellulose nitrate, cellulose acetobutyrate, rubber chloride, ethylcellulose, colophony, ester gum, benzyl abietate, cyclohexanone-formaldehyde resin, coumarone, mastic, thus, albertols, polystyrene, poly-vinyl acetate and chloroacetate. It is miscible with castor and linseed oils and with hydrocarbons. In the presence of up to 33% of alcohol it dissolves cellulose acetate, glyceryl phthalate resins, elemi, dammar. Ethyl acetate partially dissolves sandarac, kauri, manila and shellac. It does not dissolve hard copals, congo or polyvinyl chloride. It attacks polythene, Perspex and polyvinyl chloride.

n-Propyl Acetate. $CH_3 . COO . C_3H_7$

So-called 'technical' amyl acetate consists largely of n-propyl acetate along with other homologues arising from the esterification of fusel oil. Propyl acetate is seldom found in a state of purity. As is to be expected, its properties are similar to and intermediate between those of ethyl and the butyl acetates.

It is a good solvent for cellulose nitrate, ester gum, sandarac, colophony, manila, cumarone, mastic, benzyl abietate, but not for cellulose acetate or hard copals. It is miscible with castor and linseed oils and with hydrocarbons. It is partly miscible with water.

The pure ester [14] boils at 101.6° C and has sp. gr. 0.897, $0.8874_{20 4}$, $0.8822_{25 4}$; m. p. −93° C; lat. ht. 80; sp. ht. 0.47; n_D 1.3844; flash p. 57° F (14° C); s. ten. 24.6, 24.0_{25}; visc. 0.585_{20}, 0.551_{25}; elec. cond. 2.2×10^{-4}; dielec. const. 8.1; water dissolves 1.9% at 20° C. It is non-toxic and its odour is mild and pleasant.

The following binary azeotropic mixtures are known:

n-Propyl acetate (%)	(%)	b.p. (°C)
86	Water 14	82.4
60	n-Propyl alcohol 40	94.2
77	Isopropyl alcohol 23	74.8
43	Carbon tetrachloride 57	74.7

and a ternary mixture containing 19.5% of propylalcohol and 21% of water, b. p. 82.2° C.

Isopropyl Acetate. $CH_3.COO.CH\diagup^{CH_3}_{\diagdown CH_3}$

This ester is produced by esterifying isopropyl alcohol (q.v.). It has also been made by treating liquefied propylene with a mixture of acetic and sulphuric acids and similar methods. It is similar in character to n-propyl acetate, and its properties are intermediate between those of ethyl and butyl acetate. Its evaporation rate is about one-third of that f ethyl acetate.

The pure ester has sp. gr. 0.93; b. p. 88.8° C; flash p. about 46° F (8° C); m. p. $-73°$ C; n_D 1.377; v. p. 48; sp. ht. 0.52; lat. ht. 78; elec. cond. 5.7×10^{-7}; s. ten. 24.5; visc. 0.52; water dissolves 3.1% by weight at 20° C.

The following binary azeotropes are known:

Isopropyl acetate (%)	(%)	b. p. (°C))
47.7	Isopropyl alcohol 52.3	80.1
89.4	Water 10.6	76.6

British Standard Specification, BS 1834:1968, for Isopropyl acetate requires: sp. gr. 0.873–0.878 (0.869–0.874)$_{20}$; b. r. 86°–90° C, 95% min; residue 0.01% max; acidity 0.01% max. as acetic; esters 97% min; max H_2O 0.2% by weight; clearly miscible with 19 volumes of carbon disulphide at 15° C.

ASTM Specification, D657–66 for 95% grade requires: sp. gr. 0.867–0.874; b. r. 85°–90.0° C; max residue 0.005 g/100 ml; max H_2O 0.2% by weight, miscible with 19 volumes 99% heptane at 20° C; max acidity 0.01% by weight; ester value at least 95.0% by weight.

It is a good solvent for cellulose nitrate, ester gum, coumarone, elemi, sandarac, mastic, kauri, polyvinyl acetate and chloro-acetate and rubber chloride. It partly dissolves shellac, and poly-vinyl chloride, but will not dissolve cellulose acetate or hard copals. It is miscible with castor and linseed oils and with hydrocarbons.

n-Butyl Acetate. $CH_3 . COO . CH_2 . CH_2 . CH_2 . CH_3$

Butyl acetate known also as Butyl ethanoate [15] is the most widely used of all the solvents of cellulose nitrate. It is probably the best for cold lacquers and cellulose paints, since its volatility is sufficiently high for it to leave the film readily and sufficiently low to render it an excellent blush preventive, especially in conjunction with butyl alcohol; the mixture of the two effectively prevents chilling as well as gum and cotton blush.

Butyl acetate is non-toxic and has a less pronounced odour than amyl acetate; it gives solutions of cellulose nitrate of somewhat lower viscosity than the latter. In the presence of about 20% of butyl alcohol it is an excellent solvent for the less highly polymerised forms of glyceryl phthalate resin and for shellac. It is a good solvent for ester gum, benzyl abietate, cumarone, colophony, poly-styrene, poly-vinyl acetate and chloride and copolymers, cyclohexanone-formaldehyde resin, gum camphor, elemi, mastic, kauri, pontianac, sandarac, manila, rubber chloride, gutta percha resin, plextol; it partly dissolves copal ester, calcium, zinc and manganese resinates, but is a non-solvent for hard copal and cellulose acetate although it dissolves some forms of cellulose aceto-propacetate. It is miscible with castor, linseed and other oils, also with hydrocarbons. Some cellulose ethers can be dissolved in butyl acetone, notably medium viscosity ethyl cellulose.

British Standard Specification, BS 551:1965, for n-Butyl acetate requires: sp. gr. 0.884—0.887 (0.879—0.882 at 20° C, 0.876—0.879 at 25° C); b. r. 124.0°—129° C; max residue 0.01%; max H_2O 0.1%; max acids 0.01%; ester content at least 97%.

ASTM Specification, D303—58 (1965) for 90 to 92% grade requires: sp. gr. 0.874—0.876 at 20° C; b. r. 118.0°—128.0° C; max residue 0.005 g/100 ml; max H_2O 0.2%; max acids 0.01%; ester value 90.0 to 92.0%.

The following constant-boiling mixtures are known:

n-Butyl acetate (%)	(%)	b. p. ($^\circ$ C)
71.3	Water 28.7	90.2
27	n-Butyl alcohol 73	116
60	n-Propyl alcohol 40	94.2
48	Isopropyl alcohol 52	80.1

A ternary mixture, consisting of n-butyl acetate 35.3%, n-butyl alcohol 27.4%, water 37.3%, boiling at 89.4° C.

Butyl acetate dissolves water to the extent of 1.2% at 10° C, 1.28% at 15° C, 1.37% at 20° C, 1.55% at 30° C, by weight. Water dissolves 0.8% by weight of butyl acetate at 15° and 1.0% at 20° C.

Secondary Butyl Acetate. $CH_3 . COO . CH \overset{\displaystyle CH_3}{\underset{\displaystyle C_2H_5}{<}}$

Secondary butyl acetate is used to some extent in America and on the Continent, and it is made by the direct esterification of the corresponding alcohol.

Characteristics: sp. gr. 0.861, 0.872$_{20}$; b. p. 112° C; n_D1.389; v. p. 16; flash p. 64° F (17.8° C); lat. ht. 75; cub. exp. 0.00113; solubility in water 3%; dissolves 2.5% of water at 25° C. It has the fruity odour common to this class of ester; it is non-toxic. Its rate of evaporation is intermediate between that of isopropyl and n-butyl acetates. It forms a constant-boiling mixture with 86.3% of sec-butyl alcohol boiling at 99.6° C, and with 17% of water boiling at 87.4° C.

It is a solvent for cellulose nitrate, ester gum, mastic, kauri, cumarone, elemi, benzyl abietate, colophony, pontianac, manila, tar, asphalt, rubber chloride. It partly dissolves shellac and dammar. It does not dissolve cellulose acetate. It is miscible with castor and linseed oils and with hydrocarbons.

Isobutyl Acetate. $CH_3 . COO . CH_2 . CH \overset{\displaystyle CH_3}{\underset{\displaystyle CH_3}{<}}$

This ester occurs to a considerable extent in the so-called 'technical' amyl acetate along with other homologues. It is seldom used in a state of purity. It is an excellent solvent, giving solutions of low viscosity, somewhat lower than those of n-butyl acetate; these solutions, however, are somewhat prone to chilling, and the addition of amyl acetate is therefore desirable; for this reason the technical amyl acetate referred to above is preferred. Isobutyl acetate is non-toxic and its odour is mild and pleasant.

The pure ester has sp. gr. $0.8745_{20\ 4}$, $0.8695_{25\ 4}$; b. p. 117.1° C; n_D 1.3898; m. p. -99° C; flash p. 62° F $(17^{\circ}$ C); v. p. 13; lat. ht. 74; sp. ht. 0.46; s. ten. 23.7, 23.15_{25}; visc. 0.697, 0.651_{25}; elec. cond. 2.55×10^{-4}; dielec. const. 5.32; evap. period 7.7; water dissolves 0.67% at 20° C.

The following constant-boiling mixtures are known:

Isobutyl acetate (%)	(%)	b. p. ($^{\circ}$ C)
71	Water 29	90.2
55	Isobutyl alcohol 45	107.4
53	n-Butyl alcohol 47	117.2

A ternary mixture consisting of 46.5%, isobutyl acetate 23.1%, isobutyl alcohol and water 30.4%, boils at 86.8° C.

It is a good solvent for cellulose nitrate, ester gum, colophony, sandarac, benzyl abietate, coumarone, mastic, and rubber chloride, and in the presence of alcohols for glyceryl phthalate resin. It is miscible with castor and linseed oils and with hydrocarbons and partly with water. It does not dissolve cellulose acetate, copals, or copal ester.

Amyl Acetate

Amyl acetate is the doyen of the cellulose nitrate solvents, and, in spite of the many new ones that have been introduced during the last decade, still remains without equal, with the possible exception of n-butyl acetate. Its one great disadvantage is its powerful, but harmless, 'banana' or 'pear drop' odour, which to some is objectionable; its superiority over other solvents is, however, so marked that its odour is tolerated for the sake of its unique qualities.

It is made by the acetylation of amyl alcohol, obtained either from fusel oil or by synthetical methods, and is available in various grades, depending on the presence or otherwise of homologous or isomeric alcohols.

British Standard Specification, BS 552:1970 for amyl acetate requires: sp. gr. 0.872–0.880 (0.868–0.876 at 20° C); b. r. 120°–145° C, 95% min, above 135° C, 33% min; esters 95% min; acidity 0.01% max; max H_2O 0.3%; residue 0.01% max. This quality has flash p. 73° F $(22.7^{\circ}$ C)

ASTM Specifications are as follows: For amyl acetate from fusel oil (85–88% grade) D554–64: sp. gr. 0.860–0.870 at 20° C; esters 85–88%; acidity 0.03% acetic max; max residue 0.005 g/100 ml; b. r. below 110° C, none, 120° C: 15% max, 130° C: 50% max, 140° C: 60% min, above 150° C: none; max H_2O 0.2%.

For synthetic amyl acetate D318–58 (1965): sp. gr. 0.860–0.870 at 20° C; esters 85–88%; acidity 0.03% acetic max; max residue 0.005 g/100 ml; b. r. below 126° C: none, 130° C:5% max, 135° C: 25% max, 140° C: 75% min, above 155° C: none; max H_2O 0.2%.

The purest technical material available has: sp. gr. 0.876–0.878; b. r. 138°–142° C; esters 98–100%; flash p. 90° F (32° C); water dissolves 0.2%; dissolves 1% of water.

Pure iso-amyl acetate (3-methylbutyl acetate) has b. p. 139.5–140° C; lat. ht. 69; sp. ht. 0.46; n_D 1.405; visc. about 0.87; dielec. const. 4.81; cub. exp. 0.00119; m. p. -99° C; sp. gr. 0.872_{20}.

Pure n-amyl acetate [14] (pentan-1-acetate) has b. p. 149.2° C; m. p. -71° C; n_D 1.4028; sp. gr. 0.8753_{20}, 0.8707_{25}; s. ten. 25.8, 25.25_{25}; visc. $0.924, 0.862_{25}$.

The technical grade consists of the acetic esters of all the higher alcohols occurring in fusel oil, namely, n-propyl, isobutyl, isoamyl, active amyl, n-hexyl, n-heptyl alcohols; all primary alcohols. Neither the primary alcohols, isopropyl or n-butyl, nor any secondary or tertiary alcohols, seem to occur in traceable quantities [13]. The 'pure' grade consists almost entirely of isoamyl and active amyl acetates, the ratio being about four to one; the boiling-points of these two esters and of the alcohols themselves are so similar that it is not technically feasible to separate them.

There occurs in fusel oil a quantity of a high-boiling, evil-smelling substance of complex composition, which is highly deleterious to lacquer films, since it imparts its disagreeable odour to them, tends to cause them to turn brown with age, and also diminishes the toughness, strength and adhesion. Amyl acetate for lacquer work should evaporate completely, leaving no such evil-smelling residue.

The specific gravities and boiling-points of the acetates of the alcohols of fusel oil are as follows:

Amyl acetate is a good solvent for cellulose nitrate, ethyl cellulose, ester gum, colophony, dammar, albertols, benzyl abietate, cyclohexan-one-formaldehyde resin, poly-vinyl acetate and aceto-chloride, rubber chloride, gutta percha resin, cumarone, mastic, copals, kauri, sandarac, zanzibar, elemi, fluid silicones; it partly dissolves shellac, animi,

sp. gr. (15° C)	b. p. (° C)
n-Propyl acetate	0.897	101.6
Isobutyl acetate	0.875	116.3
Isoamyl acetate	0.876	138.8
Active amyl acetate	0.890	141.2
n-Hexyl acetate	0.890	169.0
n-Heptyl acetate	0.875	191.2

olibanum, carnauba wax, copal ester, linoxyn; it attacks polythene, methyl-methacrylate resin and polyvinyl chloride, it does not dissolve cellulose acetate or shellac. It is miscible with castor, linseed and other varnish oils and with hydrocarbons. When mixed with alcohols it dissolves glyceryl phthalate resins and ethyl cellulose. Pure amyl acetate is soluble in water to the extent of 0.2% at ordinary temperatures, and will dissolve 1% of water at 24° C. v. p. at 20° C, 4.5 mm. Isoamyl acetate forms a constant-boiling mixture with 97.5% of isoamyl alcohol boiling at 131.3° C, also with 45% of isobutyl alcohol boiling at 107.4° C and with 86% of sec-butyl alcohol boiling at 99.6° C.

'Synthetic amyl acetate,' produced from petroleum, and known as Pentacetate [4], consists of a mixture of five of the eight possible isomeric esters, together with small quantities of free alcohols. The ASTM specification D318–58 (1965) is given on page 000.

Its solvent properties are similar to those of fusel-oil amyl acetate, and its tolerance for toluene is about the same as that of pure amyl acetate and secondary butyl acetate, but is not so high as that of n-butyl acetate or ethyl acetate. The viscosities of its solutions of cellulose nitrate are nearly identical with those of solutions in pure amyl acetate. It forms a 'constant-boiling mixture' with 33% of water, boiling at 92–95° C.

sec-Amyl Acetate

Commercial products consisting mainly of the acetates of the secondary alcohols pentan-2-ol and pentan-3-ol are now available. They are produced by synthetical processes from petroleum distillates, and vary in quality as shown by the following figures:

a. [7], sp. gr. $0.861–0.865_{20}$; esters 89–93%; acidity 0.02% max; residue 0.005% max; b. r. 120°–140° C, 95% between 120°–135° C, 85% between 125°–135° C.

b. [8], sp. gr. 0.863_{20}; esters 85–88%; b. r. 128°–134° C; n_D

1.4021; dissolves 0.8% of water at 25° C; cub. exp. 0.00108; flash p. 89° F (31.7° C).

Sec-amyl acetate has a somewhat lower tolerance for dilution than fusel-oil amyl acetate, and its odour is slightly less provocative of coughing; its solvent properties are similar.

Hexyl Acetates

A number of isomeric hexyl acetates are available commercially; these substances are similar to amyl acetate in character but of higher boiling range. They are solvents for cellulose nitrate, polyvinyl chloride and aceto-chloride, rubber chloride, polystyrene, colophony, ester gum, dammar, kauri, coumarone, manila, elemi, pontianac, linseed oil, pitch, tar, asphalt, but not for cellulose acetate, shellac or raw rubber. Their characteristics are shown in the table opposite.

Ethyl Formate. $H . COO . C_2 H_5$

This is a highly volatile solvent, having a boiling-point similar to that of acetone; it is only used when expediency demands. Its odour is less powerful than that of acetone, and it is non-toxic. It is not a stable ester; it readily develops acidity in the presence of moisture, and it cannot therefore be recommended for metal lacquers.

Physical Characteristics; sp. gr. 0.925–0.930 (0.923_{20}); b. r. 53°–57° C; b. p. 54.3° C; m. p. 80° C; $n_D 1.360$; flash p. −2° F (−19° C); clearly miscible with benzene; soluble in 9 parts of water at 18° C; sp. ht. 0.51; lat. ht. 97; visc. 0.39; elec. cond. 1.45×10^{-9}; dielec. const. 9.1; v. p. 200.

The following azeotropic mixtures are known:

Ethyl formate 84%	Methyl alcohol 16%	b. p. 51° C
Ethyl formate 37%	Carbon disulphide 63%	b. p. 39.4° C

It is a rapid solvent for cellulose nitrate and acetate, giving solutions of low viscosity, which, however, are highly prone to chilling.

n-Butyl formate. $H . COO . CH_2 . CH_2 . CH_2 . CH_3$

Butyl formate is occasionally used, especially when it is desired to obtain a film of high strength by using a cellulose nitrate of high viscosity. It gives solutions of lower viscosity than either butyl or amyl acetate. It is a solvent for ester gum, copal ester, coumarone, benzyl

Acetate of

	Hexan-1-ol	Hexan-2-ol	4-Methylpentan-1-ol	4-Methylpentan-2-ol	2-Ethylbutan-1-ol
sp. gr.	0.8726_{20}	0.863	0.856_{20}	0.856	0.880_{20}
n	1.4096	1.408	1.394	–	1.41
b. p. (°C)	170.5	146–156	136–146	136–148	162
flash p. (°C)	–	–	13	113	135
visc. 25°	1.075	–	2.2	–	–
m. p. (°C)	–63	–	–	–	–
dil. rat:					
toluene	–	1.6	1.8	1.8	2.1
petroleum	–	0.8	1.1	1.1	–
aq. dissols (%)	–	–	0.1	0.8	0.8
diss. aq (%)	–	–	0.9	–	0.57
s. ten.	26.6	–	–	–	–

possesses the property of being a solvent for some types of cellulose acetate. It is a solvent for ester gum, copal ester, cumarone, benzyl abietate and mastic; it partly dissolves shellac, and in the presence of alcohol will dissolve glyceryl phthalate resins. It is miscible with castor, linseed and other oils and with hydrocarbons, but not with water to any considerable extent. It will not dissolve hard copals.

Pure butyl formate has sp. gr. 0.9108 (0.892_{20}); b. p. $106°-107°$ C; m. p. $-90°$ C; v. p. 23; lat. ht. 87; sp. ht. 0.46; $n_D 1.389$; visc. 0.59. It forms constant-boiling mixtures, with 16.5% of water boiling at $83.8°$ C; with 23.7% of butyl alcohol boiling at $105.8°$ C; and a ternary mixture consisting of 68.7% of ester; 10% of butyl alcohol; and 21.3% of water boiling at $83.6°$ C.

Amyl Formate. $H . COO . C_5 H_{11}$

Amyl formate is sometimes preferred to amyl acetate, especially for use in the manufacture of leather-cloth, as its odour is less pronounced and is suggestive of leather. Its volatility is very similar to that of n-butyl acetate, for which it can be substituted as required.

The 'pure' material, which consists of the formates of the isomeric amyl alcohols, has sp. gr. 0.886_{20}; b. p. $123°-124°$ C; m. p. $-73.5°$ C; $n_D 1.398$; visc. 0.8; s. ten. 25; dielec. const. 7.7 (audio freq.) Pure n-amyl formate has sp. gr. 0.885_{20}; b. p. $131°$ C; $n_D 1.400$.

Amyl formate is a good solvent for cellulose nitrate, colophony, ester gum, copal ester, benzyl abietate, coumarone, raw rubber, mastic; it is miscible with castor and linseed oils and with hydrocarbons, but not with water. It dissolves 0.3% of water at $20°$ C. It does not dissolve cellulose acetate or hard copals; it partly dissolves shellac and will dissolve glyceryl phthalate resin in the presence of an alcohol. Solutions of cellulose nitrate in amyl formate are slightly less viscous than those of corresponding concentration in n-butyl acetate.

The following binary constant-boiling mixtures are known:

(%)	(%)	b. p. ($°$ C)
Isoamyl formate 79	Water 21	90.2
n-Amyl formate 71.6	Water 28.4	91.6
Isoamyl formate 74	Isoamyl alcohol 26	123.6
n-Amyl formate 57	n-Amyl alcohol 43	130.4

n-Butyl Propionate. $CH_3 . CH_2 . COO . CH_2 . CH_2 . CH_2 . CH_3$

This ester can be used in the place of amyl acetate when a solvent of somewhat lower volatility is required, as in the case of a lacquer to be applied under excessively humid conditions. Its odour, reminding of apples, is less objectionable to most persons than is that of amyl acetate. It is non-toxic.

Butyl proprionate is manufactured in the United States from the residue arising from the manufacture of acetone by the fermentation of seaweed. This residue consists largely of calcium propionate together with small quantities of the acetate and butyrate [8].

Butyl propionate is a good solvent for cellulose nitrate, ester gum, copal ester, benzyl abietate, mastic, dammar, coumarone, elemi, rubber chloride; but does not dissolve cellulose acetate, hard copals, sandarac, zanzibar or glyceryl phthalates. It partly dissolves shellac. It is miscible with castor, linseed and other oils and with hydrocarbons, but not with water. It dissolves water to the extent of 0.452% at 7° C, 0.513% at 16° C, 0.576% at 22.5° C, and 0.691% at 30° C.

Characteristics of the pure ester: sp. gr. 0.883 (0.875_{20}); $n_D 1.401^\circ$ C; b. p. 145.5° C; flash p. 113° F (45° C); m. p. -89.6; cub. exp. 0.00108.

Amyl Propionate

Amyl propionate is of value for brushing lacquers where a rather slower rate of evaporation is required than is given by amyl acetate. Like the latter, its composition varies considerably. The following figures show the characteristics of a high-grade lacquer quality: sp. gr. 0.870–0.873; b. r. 140°–170° C; flash p. 105° F (40° C); esters 85–100%; acidity 0.02% max as propionic acid; dil. rat. toluene 1.6, petroleum 0.9. Pure iso-amyl propionate has b. p. 160° C; m. p. -73° C; lat. ht. 63; sp. ht. 0.46; $n_D 1.406$; flash p. 142° F (63° C); visc. 0.936 at 25° C; dielec. const. 4.25. Its odour is less objectionable than that of amyl acetate, being milder and more like apples. It is non-toxic. It has much the same solvent properties as amyl acetate, but is somewhat slower in its action, and gives slightly more viscous solutions. It is a solvent for cellulose nitrate, ester gum, benzyl abietate, copal ester, coumarone, elemi, mastic, sandarac, kauri and soft copals, but not for cellulose acetate, shellac, dammar or hard copals. It is miscible with castor and linseed oils and with hydrocarbons, but is not miscible with water.

Ethyl Butyrate. $CH_3 . CH_2 . CH_2 . COO . C_2H_5$

This substance is a solvent intermediate between ethyl and n-butyl acetates in solvent properties. It is a rapid solvent for cellulose nitrate, and yields solutions having viscosities somewhat lower than those of n-butyl acetate. Ethyl butyrate is a non-toxic substance having a powerful 'pineapple' odour, and is widely used for flavouring purposes.

Characteristics of Pure n-Ester: b. p. 120° C; sp. gr. 0.879_{20}; flash p. 73° F (23° C); visc. 0.67; n_D1.392.

It is a solvent for ester gum, colophony, coumarone, manila, mastic, gutta percha resin and ethyl cellulose, but not for cellulose acetate or glyceryl phthalate resin. It attacks perspex, polythene and polyvinyl chloride but not nylon.

n-Butyl Butyrate. $CH_3 . CH_2 . CH_2 . COO . C_4H_9$

In recent years this ester has been used as a cellulose nitrate solvent to a small extent. It consists usually of a mixture of the n-butyl esters of n- and iso-butyric acids, the former preponderating. Pure n-butyl n-butyrate has sp. gr. 0.869_{20}; b. p. 165° C; n_D1.406; visc. 0.97_{25}.

It is a solvent for cellulose nitrate, ester gum, mastic, coumarone, dammar, elemi, shellac, zinc resinate; but not for cellulose acetate, sandarac, zanzibar, hard copals. n-Butyl butyrate dissolves water to the extent of 0.380% at 10° C, 0.424% at 16° C, 0.497% at 25° C, 0.574% at 33° C. Its odour is strong and reminiscent of apples; it is non-toxic.

n-Butyl butyrate is useful when it is desired to have a solvent that evaporates somewhat more slowly than amyl acetate. It imparts good brushing flow and gloss to lacquers.

Benzyl Formate. $C_6H_5 . CH_2 . OOCH$

Benzyl formate is similar in character to benzyl acetate, but is of somewhat greater volatility. Its odour is less pronounced and is suggestive of leather. It is not widely used.

Physical Characteristics: sp. gr. 1.08–1.09; n_D 1.519–1.520; b. r. 200°–202° C.

It is a good solvent for cellulose acetate and nitrate, for ester gum, copal ester, benzyl abietate, coumarone, glyceryl phthalate resins, and is miscible with castor and linseed oils, and with aromatic and petroleum hydrocarbons.

Benzyl Acetate. $C_6H_5 . CH_2 . OOC . CH_3$

Prepared by the direct esterification of benzyl alcohol with acetic acid or by the action of anhydrous sodium acetate on benzyl chloride. It is an excellent solvent of medium volatility for cellulose nitrate and for some forms of cellulose acetate. Its odour is powerful but pleasant, and is suggestive of jasmine flowers.

Physical Characteristics: sp. gr. 1.060–1.062; n_D 1.502–1.503; b. r. 215°–216° C; m. p. –51.5° C; flash p. 216° F (102° C); esters 99–100%; non-toxic; visc. 1.425 at 25° C.

It yields solutions, of moderate viscosity, of cellulose acetate and nitrate, and has a temporary softening effect on films made of these materials. It imparts good brushing flow and high gloss. It is a good solvent for ester gum, copal ester, benzyl abietate, coumarone, mastic, castor and linseed oils, and in conjunction with alcohols it dissolves glyceryl phthalate resins and shellac; it will dissolve copals on prolonged heating. It is miscible with aromatic and petroleum hydrocarbons and all the usual solvents.

Methyl Lactate. $CH_3 . CH(OH) . COO . CH_3$

Methyl lactate is a colourless liquid of mild odour, it is a solvent for cellulose nitrate, acetate, aceto-propionate, aceto-butyrate and cellulose ethers. It has a high tolerance for diluents usually about 20% higher than ethyl lactate while its boiling point is 10° C lower.

It has sp. gr. 1.09_{20}; b. p. 144.8 $\triangle^-$ n_D 1.413; m. p. about –66° C; flash p. 61° C; ht. comb. 4778; It is completely miscible with water and most solvents.

Ethyl Lactate. $CH_3 . CH(OH) . COO . C_2H_5$

Ethyl lactate, known also as ethyl 2-hydroxy propanoate [17], is one of the most important of the high-boiling solvents for both cellulose nitrate and acetate; it has exceptionally high solvent powers, tolerating considerable dilution with non-solvents. Its rate of evaporation is slower than is desirable but this slowness is of advantage where brushing lacquers are concerned, or where the lacquer has to be applied in a cold, humid atmosphere; it also favours the production of films with a high gloss, but films produced by means of this vehicle have a tendency to remain soft for a considerable time beyond what might be considered to be a normal drying period. It imparts good flowing properties to lacquers, allowing them to be applied in thin coats over widely varying types of surface with the production of smooth, impermeable, and uniform films.

Compared with amyl and butyl acetates, its solvent action on cellulose nitrate is distinctly slower and its solutions are of decidedly higher viscosity, but they will tolerate two to three times as much non-solvent diluent; quantities of water—up to 25%— can be added without causing precipitation or blushing on subsequent evaporation. The dilution latitude of solutions of cellulose acetate in ethyl lactate is two to twelve times as great as those of other solvents.

Ethyl lactate is manufactured by the esterification of lactic acid obtained from the fermentation of sugar solutions with *Bacillus acidi lactici*. It has also been made by a purely synthetical method [9] as follows: Acetaldehyde is first obtained, either by the hydration of acetylene or by the dehydrogenation of alcohol, and is combined with hydro-cyanic acid to form lactonitrile (acetaldehyde cyanhydrine). This nitrile is converted directly into ethyl lactate by treatment with alcohol and hydrochloric or sulphuric acid, ammonium chloride or sulphate being formed simultaneously. The ammonium salt is separated by mechanical means, and the ester is purified by fractional vacuum distillation; yields of 90% are said to be obtained.

Ethyl lactate exists in two stereoisomeric modifications and may contain lactides or inner esters of lactic acid; these are solvents with higher boiling-points than ethyl lactate, and cause a widening of the boiling-range and of the limits for the other physical properties.

British Standard Specification, BS 663—1957 for ethyl lactate requires: sp. gr. 1.036—1.042, (1.032—1.038 at 20° C); b. r. 145°—160° C, 95% min; residue 0.01% max; acidity 0.025% max as lactic acid; esters 97% min; max H_2O 0.70%.

Pure ethyl lactate has sp. gr. 1.031_{20}; b. p. 154.5° C; n_D 1.4118; flash p. 115°F (46°C). v. p. 1.8; cub. exp. 0.00098; m. p. −25°; visc. 2.6 at 25° C; s. ten. 30.

Ethyl lactate, when pure, is nearly odourless; technical products have a weak odour resembling that of ethyl butyrate, and should not leave a disagreeable odour on evaporation.

Ethyl lactate is a solvent for cellulose acetate, nitrate, aceto propionate and acetobutyrate, also for benzyl and ethyl cellulose, zein, colophony, manila, kauri, pontianac, shellac, benzyl abietate, thiourea and urea-formaldehyde resins, glyceryl phthalate resins, polystyrene, polyvinyl acetate, basic dyes; and is miscible with raw linseed oil, blown castor oil, aromatic hydrocarbons, low-boiling hydrocarbons, white spirit and water. It will dissolve hard copals on prolonged boiling. It will not dissolve rubber chloride and is without action on vulcanised rubber.

In conjunction with phthalic anhydride it is used for producing frosting effects in lacquers.

Isopropyl Lactate. $CH_3.CH(OH).COO.CH \begin{smallmatrix} CH_3 \\ \\ CH_3 \end{smallmatrix}$

This ester, like ethyl lactate, is a solvent for both cellulose acetate and nitrate; it is better solvent for ester gum than is ethyl lactate. It also dissolves colophony, shellac, benzyl abietate, coumarone, glyceryl phthalate and vinyl acetate resins. Technical preparations have sp. gr. 0.994–0.998; b. r. $155°-165°$ C (95%); esters 95% min; acidity 0.05% max, as lactic acid.

Pure material has sp. gr. 0.998_{20}; $n_D 1.408$; b. p. $166-8°$ C.

Butyl Lactate. $CH_3 . CH(OH) . COO . C_4H_9$

n-Butyl lactate is made by the direct esterification of lactic acid with butyl alcohol. Like other esters of lactic acid, it exists in two stereoisomeric modifications. When pure, it is a colourless liquid, but the technical article is often slightly brown. It is practically devoid of odour, and is non-toxic.

Physical Characteristics; sp. gr. 0.984–0.988; b. r. $185°-195°$ C; esters 95% min; acidity 0.05% max.

Pure material has sp. gr. 0.973_{20}; $n_D 1.422$; b. p. $185°$ C; lat. ht. 77; m. p. $-43°$ C; flash p. $81°$ C; v. p. 0.4; evap. period 433. Water dissolves 3.4% at $25°$ C; Butyl lactate dissolves 13% of water at $25°$ C, but it is not hygroscopic.

The commercial product varies rather widely in quality, owing to the presence of condensation products such as lactide.

It is a good solvent for ester gum, and renders this resin compatible to some extent with cellulose acetate; it readily dissolves cellulose nitrate, ethyl cellulose, zein, glyceral phthalate resin, coumarone; it has a moderate solvent action on cellulose acetate, rubber chloride, copal ester, benzyl abietate, shellac, mastic. It is miscible with castor and linseed oils, and with aromatic and petroleum hydrocarbons, but not with water. It has a high tolerance for diluents.

It imparts brilliance and tenacity to films, but tends to cause softness for a considerable period owing to its slow rate of evaporation.

Butyl lactate tends to prevent the formation of a skin on oil varnishes and paints.

Isobutyl Lactate. $CH_3.CH(OH).COO.CH_2.CH\begin{smallmatrix}\diagup CH_3 \\ \diagdown CH_3\end{smallmatrix}$

Isobutyl lactate is similar to n-butyl lactate in solvent properties, but it evaporates somewhat more rapidly. This ester has b. r. $168°-200°$ C; sp. gr. 0.974–0.978; acidity 0.05% max, as lactic acid; esters 95% min. Pure material has sp. gr. 0.97_{20}; $n_D 1.418_{25}$.

sec-Butyl Lactate. $CH_3.CH(OH).COO.CH\begin{smallmatrix}\diagup CH_2CH_3 \\ \diagdown CH_3\end{smallmatrix}$

Secondary butyl lactate [10] is an ester of similar character to the other isomers and has sp. gr. 0.974_{20}; b. p. $180°$ C. It has a faint fruity odour, and is but slightly soluble in water.

Amyl Lactate. $CH_3 . CH(OH) . COO . C_5H_{11}$

This ester is used as a softener or as a high-boiling solvent when it is desirable to prolong the period of drying of a cellulose-nitrate lacquer. It is not widely employed.

Physical Characteristics: sp gr. 0.968–0.972; b. p. about $210°$ C; n_D 1.424_{25}; acidity 0.05% max. Like other esters of lactic acid, the technical product is apt to vary considerably in composition, and to contain lactides.

It is a solvent for cellulose nitrate, ester gum, copal ester, benzyl abietate, shellac, cumarone and mastic. It does not dissolve cellulose acetate, nor hard copals, although the latter can usually be dissolved on prolonged boiling. It dissolves glyceryl phthalate resins in conjunction with an alcohol. It is miscible with castor and linseed oils and with hydrocarbons, but not with water.

Its odour is weak, somewhat like that of brandy; it is non-toxic.

Ethyl Benzoate. $C_6H_5 . COO . C_2H_5$

This is a solvent of the 'high-boiling' class, and suffers from the defects inherent in this class. It is useful for imparting good brushing properties to cellulose lacquers, and gives films of high gloss, but these films tend to remain soft for an unduly long period; it can be used as a plasticiser.

Characteristics: sp. gr. 1.051–1.053, (1.047_{20}); b. p. 213° C; n_D 1.5052; m. p. −34° C; visc. 2.2; elec. cond. 1×10^{-9}. dielec. const. 6; cub. exp. 0.00092.

The technical article is usually one of high purity. Ethyl benzoate is a solvent for cellulose acetate and nitrate, ethyl cellulose, ester gum, sandarac, mastic, shellac, cumarone, glyceryl phthalate resin, benzyl abietate, but not for hard copals, except on prolonged boiling. It is miscible with castor and linseed oils and with hydrocarbons, but not with water.

Its odour is powerful but pleasant; it is non-toxic and is stable to light.

Methyl Benzoate

Methyl benzoate, known also as 'Oil of Niobe,' has a somewhat less pronounced odour than the ethyl ester.

Characteristics: sp. gr. 1.093–1.094 (1.089_{20}); b. p. 199; n_D 1.517; elec. cond. 1.37×10^{-3}; diel. const. 6.58; visc. 2.07.

It is a solvent for cellulose acetate and nitrate, ester gum, cumarone, rubber. It is miscible with oil, but not with water. Non-toxic.

Diethyl Carbonate. $(C_2H_5O)_2 . CO$

This is a colourless liquid, having a weak odour resembling that of ethyl butyrate, with a faintly lachrymatory after-effect similar to that of ethyl chloroacetate probably due to faint traces of impurities.

It is manufactured by treating phosgene (carbonyl chloride) with alcohol, either by refluxing them together and extracting the ester with a suitable solvent, such as carbon tetrachloride, or by treating alcohol in excess with phosgene vapour, [10] and in other similar ways. Phosgene is produced from producer gas and chlorine by passing them over activated carbon at 125°–150° C.

Diethyl carbonate is also produced when ethyl chloroformate and alcohol are treated with di-methyl aniline.

Physical Characteristics: sp. gr. 0.975; n_D 1.385; b. p. 127° C; b. r. 120°–130° C; v. p. about 30; flash p. 86° F (30° C); lat. ht. 73; sp. ht. 0.46; evap. period. 14; elec. cond. 17×10^{-9}; dielec. const. 3.15; cub. exp. 0.0012; m. p. −43° C; dil. rat. toluene 0.6, petroleum 0.4; it dissolves 1.4% of water at 25° C.

Diethyl carbonate is miscible with castor oil, aromatic hydrocarbons, and most organic solvents, but non-miscible with some petroleum hydrocarbons. It dissolves 2.6% of water at 20° C. It is not, by itself, a good solvent for cellulose acetate or nitrate, but the addition of an alcohol or an ester greatly increases its solvent power for the latter, and also renders it a solvent for benzyl cellulose.

Its solutions of cellulose nitrate are more viscous than solutions of corresponding concentration in butyl acetate, and these solutions have but a poor tolerance for hydrocarbons; the presence of a small proportion of ethyl lactate considerably increases the tolerance.

Diethyl carbonate is a solvent for ester gum, cumarone, mastic, polyvinyl acetate, colophony and thus; the presence of 5 to 10% of anhydrous alcohol renders it a solvent for elemi, sandarac, manila, kauri. It does not dissolve shellac, copal, congo, copal ester, glyceryl phthalate resin, benzyl resin or polyacrylonitrile.

Dimethyl Formamide $H.CO.N \begin{smallmatrix} CH_3 \\ \\ CH_3 \end{smallmatrix}$

This product has achieved some acceptance as a high boiling ingredient of inks and surface coatings—especially those for application to PVC. It is also employed in the spinning of acrylic and polyurethane fibres, as a crystallization and reaction solvent, in the pharmaceuticals industry, as a dyestuffs solvent, as an active ingredient in heavy-duty cleaners and paint strippers, and for extracting acetylene and fritadiene from process gas streams, in the manufacture of synthetic leathers. Its solvent properties are:

Soluble	*Insoluble*
Polyacrylonitrile	Polyethylene
Polyurethanes	Polypropylene
Polymethylmethacrylate	Polytetrafluoroethylene
Cellulose acetate	Saturated polyesters
Cellulose nitrate	Urea-formaldehyde resins
Cellulose acetate butyrate	Natural rubber
Ethylcellulose	Butyl rubber
Cyanoethylated cellulose	Styrene-butadiene rubber
Polystyrene	Nylon 66, 6, and 610.
Polyvinyl chloride	
Polyvinyl alcohol	
Polyvinyl acetate	
Alkyds	
Phenol-formaldehyde resins	
Coumarone-indene resins	
Shellac	
Ester gum	
Kauri gum	

The odour is not unduly powerful and some evidence of cumulative systemic injury to animals exposed to dimethyl formamide has been recorded. The American Conference of Governmental Industrial Hygienists 1969 recommended 10 ppm as the safe working concentration for continuous exposure. Dimethyl formamide has also been found to react violently with certain other chemicals when heated, such as aluminium trimethyl cyannic chloride, sodium metal, and magnesium nitrate.

Pure dimethyl formamide has the following properties: sp. gr. 0.953; n_D 1.4306; visc. 0.802_{25}; flash p. (Pensky Marten) 135° F (57° C); dielec. const. 36.71_{25}; v. p. 3.7; b. p. 153; evap. rate (butyl acetate = 100) 17;

References

[1] U.S. Pat. 1467091 ; 1467105.
[2] Brit. Pat. 131088.
[3] *Cf.* W. S. Simmons. "Alcohol," Macmillan & Co. Ltd.
[4] Airco Export Corp., New York, U.S.A., Sharples Chemicals Inc.
[5] Asiatic Petroleum Co. Ltd., London.
[6] Park and Hopkins. *Ind. Eng. Chem.*, 1930, p. 826.
[7] Holzverkohlungs Industrie, Konstanz.
[8] *Cf.* B.P. 396968.
[9] Canadian Electro-Products Company, Shawinigan; American Cyanamide Co., N.Y., U.S.A.; Brit. Pat. 257907.
[10] *Cf.* U.S. Pat. 1956972.
[11] American Cyanamide Company, N.Y., U.S.A.
[12] *Cf.* U.S. Pat. 1926510 ; 1926511.
[13] U.S. Pat. 1603703 ; 1638014 ; 1603689.
[14] *Cf. J.C.S.*, 1950, p. 79.
[15] International Union of Chemistry.

CHAPTER FIVE

Glycols and their ethers

Ethylene glycol. $HO.CH_2.CH_2.OH$

Ethylene glycol, known also as ethane 1,2-diol, is a liquid with properties intermediate between those of ethyl alcohol and glycerin. It is colourless and odourless, has a bitter-sweet taste, somewhat viscous, very hygroscopic, and miscible in all proportions with water, acetone, methyl ethyl ketone, cyclohexanone, alcohol, butanol and furfural, non-miscible with vegetable oils, turpentine, decalin, tetralin, hydrocarbons, ether, carbon tetrachloride, carbon disulphide, esters or chloroform. b. p. 197.6° C; sp. gr. 1.117; flash p. 240° F (115° C); ign. p. 120° C; auto-ignit. temp. 417° C; m. p. about −12° C; sp. ht. 0.575; $n_D1.431$; lat. ht. 191; s. ten. at 25°C 53; visc. (see p. 173); v. p. 0.06; cub. exp. 0.00062; elec. cond. 1.07×10^{-6}; dielec. const. 41.2 at 25°C.

British Standard Specification BS 2537:1955 for ethylene glycol requires: completely miscible with water at 15° C; sp. gr. 1.116–1.118 $(1.114–1.116)_{20}$; b. r. 194°–199° C 95% min.; ash 0.01% max; acidity 0.01% max; Not alkaline; Ionisable and hydrolysable chlorine 0.01% max; free from sulphates.

The specific gravities and crystallising points of mixtures of glycol and water are approximately as follows:

Per cent. Glycol	sp. gr.	° F.	° C.
25	1.038	+10	−12
30	1.045	+3	−16
35	1.052	−5	−20
40	1.058	−15	−26
45	1.064	−26	−32

The freezing points have also been quoted as follows [17] :

Glycol%	sol. p. ° C.		
	1	2	3
10	− 3.7	− 2.9	− 4
20	− 8.4	− 9.7	− 9
30	−14.9	−17.6	−16
40	−24.0	−26.0	−
50	−36.1	−37.0	−

There is a eutectic containing 60% Glycol sol. p. —49° C.

It is a solvent for casein, gelatine, dextrine, low viscosity silicones, some phenol-formaldehyde resins and dyes, and zein, and gelatinises cellulose nitrate. It does not dissolve cellulose ethers or esters, waxes, kauri, dammar, ester gum, vinyl resins, rubber or rubber chloride. It partly dissolves shellac and rosin. It plasticises cellophane, parchment, tracing paper. The method of manufacture is by liquid phase hydration of ethylene oxide.

The toxicity of glycol is in question; when taken internally it appears to form the poisonous substance oxalic acid.

Ethylene glycol monoacetate
$CH_3 . COO . CH_2 . CH_2 . OH$

This is an odourless, colourless liquid, miscible with water and with aromatic hydrocarbons, but non-miscible with paraffins or linseed oil. b. r. 185°–195° C; sp. gr. 1.11; flash p. 215° F (102° C). It is a high-boiling solvent for cellulose nitrateand acetate, elemi and colophony. It partly dissolves mastic, but not shellac, kauri, sandarac, dammar, zanzibar, hard copals, ester gum, and coumarone

Ethylene glycol diacetate
$CH_3 . COO . CH_2 . CH_2 . OOC . CH_3$

The diacetate sometimes called ethylene diacetate is made by heating ethylene dichloride with potassium or sodium acetate in the presence of about 5% of glycol, which acts as a flux [1] under a pressure of 8 to 10 atm.

It is a colourless liquid having a slight odour, reminiscent of ethyl acetate. sp. gr. 1.15; b. p. 190; n_D 1.415; flash p. 220° F (105° C); v. p. 0.3; elec. cond. 2.8×10^{-2}.

It is a solvent for cellulose acetate and nitrate, ethyl cellulose, benzyl cellulose, mastic, colophony, and gum camphor, but not for shellac, kauri, sandarac, dammar, zanzibar, hard copal, ester gum, coumarone. It is not miscible with petroleum or linseed oil, and its solutions of cellulose esters have a low tolerance for toluene and xylene of about 1.4.

Ethylene glycol monomethyl ether
$CH_3O . CH_2 . CH_2 . OH$

Known also as methyl glycol, 2-methoxyethanol, methyl cellosolve [2], and 'methyl oxitol' [20], it is the only ether of this group which is a solvent for low viscosity cellulose acetate as well as for the nitrate. It also dissolves poly-vinyl acetate and chloro acetate, vinyl acetals, zein, shellac, kauri, mastic, sandarac, elemi and colophony, and to some extent ester gum, zanzibar and dammar. It is miscible with aromatic and light paraffin hydrocarbons,and also with water, but not with heavy paraffins or linseed oil. It is without action on polyvinyl chloride, raw rubber. It is a colourless liquid of mild odour, and is a solvent for spirit-soluble dyes. sp. gr. 0.9748; b. p. 124.5° C; flash p. 107° F (42° C); auto-ignition temp. 288° C; n_D 1.4023; visc. 1.7; s. ten. 35; v. p. 7.5; sp. ht. 0.534; lat. ht. 135; elec. cond. 1.1×10^{-6}. It is practically odourless; under certain conditions it decomposes rapidly into acetaldehyde and methanol.

The following azeotropes are known:

		b. p. (° C)
Methyl glycol 24.7%	Water 75.3%	100
Methyl glycol 15%	Cyclohexane 85%	77.5
Methyl glycol 9%,	Cyclohexane 52%, Benzene 39%.	73.0

Methyl glycol forms mixtures with water having specific gravities over 1 at 15° C in the range of 20–80% of water with a maximum of 1.014 at 60%.

Ethylene glycol monomethyl ether acetate
$CH_3 . COO . CH_2 . CH_2 . OCH_3$

Methoxy ethyl acetate known commercially as methyl glycol acetate [2] or methyl oxitol acetate [20] is a colourless liquid having a pleasant weak odour. It is miscible with water in all proportions as well as with most organic liquids and solvents; it forms a number of binary azeotropes with the latter. Its dilution ratio with toluene is 2.3 (nitrocellulose); sp. gr. 1.0067; b. p. 145.1° C; flash p. (Cleveland open cup) 140° F (60° C); n_D 1.4019; ester content 95%; acidity as acetic acid 0.02%.

It is a solvent for cellulose nitrate, acetate and acetate-butyrate, and ethyl cellulose as well as for polyvinyl acetate and aceto chloride, soluble vinyl copolymers, rosin,estergums, synthetic resins, including epoxides, polystyrene, and some polyurethane resins. It does not affect rubber. It imparts good flow and anti-brush characteristics on account of its high solvent powers and medium evaporation rate. Its toxicity and dermaticity are low and reasonable precautions are all that need to be taken in handling it.

Ethylene glycol monoethyl ether
C$_2$H$_5$O . CH$_2$. CH$_2$. OH

The monoethyl ether known also as ethyl glycol, 2-ethoxyethanol, 'cellosolve' [2] and 'oxitol' [20], is the most important solvent of this class. It is a colourless, nearly odourless, stable liquid, and is a good solvent for cellulose nitrate. It has the property of developing a haze when clear solutions of certain resins are mixed with clear solutions of cellulose nitrate. Ordinary ester gum, although soluble in the ether, is not compatible with a solution of cellulose nitrate, but an addition of alcohol-soluble dewaxed dammar ensures compatibility, provided that the proportion of ester gum is not more than three times that of the dammar. Small additions of glyceryl phthalate resin also render ester gum compatible. The addition of other solvents will often remove the haze, benzyl alcohol, amyl acetate and ethyl lactate being suitable for this purpose. In the absence of resins, solutions of cellulose nitrate in ethylene glycol monoethyl ether have a high tolerance for dilution with non-solvents; thus for toluene it varies from 5.6 to 6.6, for butyl alcohol it is about 7.5, and for gasoline 1.2 to 1.5, depending on the particular quality of the cellulose nitrate, and the glycol ether is therefore similar to ethyl lactate in this respect.

It has been recorded [3] that, like acetone and ethyl acetate, the methyl and ethyl ethers of ethylene glycol sometimes yield opaque films of cellulose nitrate, especially when the drying operation is conducted in a highly humid atmosphere. This may be partly due to the fact that these ethers are powerful solvents for water. Notwithstanding these defects, the ethers offer the advantage over most other solvents of similar boiling-point in having but a slight odour.

Ethylene glycol monoethyl ether can be manufactured by several methods, the first step in the process being the preparation of ethylene chlorhydrine (q.v.). From this point the process may proceed by several routes; thus one method is to transform the chlorhydrine to ethylene glycol [4] by passing the constant-boiling mixture, resulting from the distillation of the ethylene chlorhydrine solution in water, in vapour form up a tower, down which sodium carbonate solution is falling; hydrolysis takes place and the glycol, dissolved in the solution of salt, which is simultaneously formed, passes to the bottom of the tower; carbon dioxide, acetaldehyde, steam and other volatile substances passing away from the top. The ethylene glycol is subsequently recovered from the salt solution by fractional distillation. The ehtyl ether of the glycol can be made by refluxing for three hours a mixture of two molecular proportions of the glycol with two of sodium hydroxide and one of diethyl sulphate, and then fractionally distilling the mixture under diminished pressure; a yield of about 60% is obtained [5]. The ether is also made from ethylene chlorhydrine by first converting this into ethylene oxide and adding this, while stirring, to cold ethyl alcohol, containing 1% of sulphuric acid, the product being neutralised and fractionally distilled. Other ethers of ethylene glycol may be made in a similar manner by the use of other alcohols or phenols, such as methyl, propyl, or butyl alcohol; yields of 90% are obtained [6]. Catalysts other than sulphuric acid may also be used [7]. The conversion of ethylene chlorhydrine into the ethylene oxide required for the above process can be effected by passing the chlorhydrine at high temperatures over solid bases or through suspensions of lime or soda-lime in water [8] : solutions as dilute as 8% of ethylene chlorhydrine can be used for this process, and strong alkalis may also be employed. Ethylene oxide is also produced by the limited oxidation of ethylene by treatment with oxygen under high pressure at $150°-400°$ C in the presence of a silver catalyst [9, 13, 19], excessive oxidation being inhibited by adding controlling agents such as ethylene dichloride, xylene, glycol ethers, sulphur compounds or arylamines [18].

British Standard Specification, BS 2713:1966 for ethylene glycol monoethyl ether requires: sp. gr. $0.934-0.937$ ($0.931-0.934_{20}$, $0.928-0.931_{25}$); b. r. $133.5°-136.5°$ C; max. H_2O 0.1%; not alkaline; max. acid 50 ppm; miscible with distilled H_2O.

ASTM Specification, D331−66 for ethylene glycol monoethyl ether requires: sp. gr. $0.929-0.932$ at $20°$C; b. r. $132°-136°$ C; max. residue 0.005g/100ml; max. H_2O 0.3%; max. acidity 0.01%.

The pure material has sp. gr. 0.936, 0.931_{20}; b. p. $135.1°$ C; n_D 1.408; m. p. below $-70°$ C; cub. exp. 0.00097; visc. 1.7; v. p. 38; flash p. $104°$ F ($40°$ C); sp. ht. 0.555; s. ten. 32 at $25°$ C; elec. cond. 6.88×10^{-6}; auto-ignition temp. $238°$ C. It is miscible with water in all proportions, also with hydrocarbons and castor oil. It is a solvent for cellulose nitrate, but not for the acetate or propionate, benzyl cellulose, polyvinyl chloride or chloracetate. It dissolves low viscosity silicones, colophony, kauri, dewaxed dammar, glyceryl phthalate resin, benzyl abietate, zein, polyvinyl acetate and cyclohexanone-formaldehyde resin. It partly dissolves ester gum, shellac, mastic, and coumarone, and will dissolve hard copals on prolonged heating. It will not dissolve copal ester, rubber chloride or raw rubber. Exposure to sunlight tends to cause decomposition.

Ethylene glycol diethyl ether
$C_2H_5O.CH_2.CH_2.OC_2H_5$

Ethylene glycol diethyl ether is a colourless liquid having a faint odour. b. p. 121.4° C; sp. gr. 0.853; n_D 1.3914; v. p. 9.4; flash p. 95° F (35° C). It dissolves ester gum, elemi, zanzibar, coumarone, dammar, mastic, kauri, shellac, zein and oils, but not cellulose acetate or nitrate, polyvinyl resins, hard copal or sandarac. At 20° C it dissolves 3.4% of water and water dissolves 1%.

2-Ethoxyethyl acetate
$C_2H_5O.CH_2.CH_2.OOC.CH_3$

This is a colourless liquid with a weak, pleasant, ester-like odour, known commercially as 'cellosolve acetate.' [2] and 'oxitol acetate'. It is a solvent for cellulose nitrate, giving viscous solutions,nd its dilution ratios for solutions of this under standard conditions are for toluene 3.0, gasoline 1.5, turpentine 2.8.

ASTM Specification, D 343—64 for 95% grade requires: sp. gr. 0.971—0.976 at 20° C; b. r. up to 145° C: none, 150°—160° C: 90% min, above 165° C: none. Esters 95% min; acidity 0.02%.

The material also has b. p. 156° C; s. ten. 31.8; visc. 1.2 at 25° C; sp. ht. 0.494; v. p. 2; elec. cond. 2×10^{-8}; flash p. 124° F (51° C); n_D at 25° C about 1.405; water dissolves 22% at 20° C and it dissolves 6.5% of water.

It is a solvent for cellulose nitrate, ethyl cellulose, colophony, ester gum, polystyrene, polyvinyl acetate and chloroacetate, coumarone, mastic, kauri, shellac, gutta percha resin, rubber chloride, hydrocarbons and vegetable oils. It partly dissolves elemi and sandarac, but not polyvinyl chloride, raw rubber, zanzibar, or hard copal. It dissolves low viscosity cellulose acetate.

Ethylene glycol mono-n-butyl ether.
$C_4H_9O.CH_2.CH_2.OH$

The monobutyl ether is a high-boiling solvent known also as butyl glycol, and commercially as 'butylcellosolve' [12] or 'butyl oxitol'. It is a good solvent for cellulose nitrate, but rather slow in its action. Its dilution ratio under the standard condition is 3.5 for toluene, 2.3 for gasoline, and 0.375 for water.

It is a solvent for colophony, shellac, sandarac, mastic, kauri, gutta percha resin, coumarone, and elemi. It is miscible with water, linseed oil and with hydrocarbons. It partly dissolves ester gum, dammar and zanzibar, but not copal, cellulose acetate or raw rubber.

ASTM Specification, D330—64 for ethylene glycol mono-n-butyl ether requires: sp. gr. 0.900—0.905 at $20°$ C; b. r. $166°-173°$ C; acidity 0.01% acetic max; $H_2 O$ 0.3% max.

It is a colourless liquid of mild odour. sp. gr. 0.915, 0.902_{20}; n_D 1.418; b. p. $171°$ C; flash p. $140°$ F $(60°$ C); auto-ignition temp. $244°$ C; visc. 3.3; sp. ht. 583; s. ten. 31.5 at $25°$ C; elec. cond. 0.432×10^{-6}.

It is recommended for brushing lacquers and for applying second coats over existing coats of cellulose nitrate, also for reducing the viscosity of lacquers.

Propylene glycol. $CH_3 . CH(OH) . CH_2 . OH$

This glycol, propan-1,2-diol, is a colourless, somewhat viscous, bitter tasting, non-poisonous liquid, hygroscopic and soluble in water. It has sp. gr. 1.308_{20}; vise. 56.0; b. p. $188°$ C; n_D 1.432 at $25°$ C; flash p. $220°$ F $(103°$ C); v. p. 0.08; sp. ht. 0.59; lat. ht. 169; cub. exp. 0.000695; s. ten. 40 at $25°$ C. It dissolves colophony, zein, saponin and partly dissolves shellac and kauri but not cellulose esters, polyvinyl chloride, acetate or chloracetate, ester gum, dammar, rubber, hydrocarbons or vegetable oils. The diacetate has b. r. $182-184°$ C.

Butylene glycol. $CH_3 . CH(OH) . CH_2 . CH_2 . OH$

1,3 Butylene glycol is prepared by hydrogenating aldol [17]. It is a syrupy, colourless, odourless liquid similar in character to ethylene glycol.

It has b. r. $204°-208°$ C; b. p. (pure) $207°$ C; sp. gr. 1.01—1.03; n_D 1.442; flash p. $104°$ F $(40°$ C). It is soluble in water, alcohols, ethyl acetate, acetone, but not in castor oil, hydrocarbons or ether. It is a solvent for shellac and colophony, and partly for dammar and coumarone; it does not dissolve cellulose esters or ethers or ester gum, vinyl resin and rubber chloride, it plasticises rubber latex.

The *monoethyl ether* has b. p. $164°$ C; sp. gr. 0.886; n_D 1.4162.

Butylene Glycol Diacetate

Butylene glycol diacetate is a powerful solvent for both cellulose esters and ethers, also for colophony, ester gum, alkyld resins, coumarone, vinyl acetate. It partly dissolves shellac, mastic, congo, manila and rubber chloride, but not dammar, sandarac or vulcanised rubber.

It is a colourless, nearly odourless, bitter-tasting liquid, having sp. gr. 1.026–1.034; b. r. 204–210 (90%); b. p. 208° C; n_D 1.42; dil. rat. toluene 2.3; it is stable to light and probably non-toxic; water dissolves 4.8% at 20° C.

Diethylene glycol. $HO . CH_2 . CH_2 . O . CH_2 . CH_2 . OH$

Diethylene glycol, or, as it is more correctly named, sym. dihydroxy-diethyl ether, is produced by the partial dehydration of glycol and, more readily, by the combination of ethylene oxide and glycol; the former is produced by passing ethylene chlorhydrine over solid bases or through suspensions of lime or soda lime at elevated temperatures [10] . Triethylene and tetraethylene glycols are obtained as by-products.

Physical Characteristics: Diethylene glycol is a colourless, almost odourless liquid, considerably more viscous and more hygroscopic than ethylene glycol. Rinkenbach [11] gives the following figures for the comparative viscosity of the mono- and diethylene glycols:

Temperature	Diethylene Glycol	Ethylene Glycol
° C.	centipoises	centipoises
15	50	26
17.5	44	23
20.0	38	21
22.5	33	19
25.0	30	17
27.0	27	16

sp. gr. 1.121, 1.118_{20}, sol. p. −9° C; flash p. 275° F (135° C); n_D 1.447; lat. ht. 129; ign. p. about 130° C; auto-ign. temp. 229° C; b. p. 244.5; v. p. less than 0.01; sp. ht. 0.50; s. ten. 48; visc. 38; cub. exp. 0.00064; elec. cond. 0.586×10^{-6}. Miscible with water, alcohols, glycols, acetone, furfural, glycol ethers and esters, cycloxexanone, methylene dichloride, ethylene dichloride, chloroform, some esters.

A 40% solution in water has sol. p. −18° C and a 50% sol. p. −28° C.

Diethylene glycol is non-miscible with ether, benzene, toluene, decalin, tetralin, carbon tetrachloride, linseed oil, castor oil, petroleums. It dissolves cellulose nitrate, colophony, and dyes, but is a non-solvent for ester gum, coumarone, copals, kauri, dammar, rubber, cellulose ethers, polythene, polyvinyl chloride, rubber chloride and cellulose acetate; mastic is partly soluble. It is a good plasticiser for cellophane and for shellac in proportions up to 25%.

Diethylene glycol monoacetate is a solvent for cellulose acetate and nitrate, colophony, and gum camphor. It partly dissolves coumarone, mastic, elemi, and kauri, but not ester gum, shellac, copals, sandarac. It is miscible with water and aromatic hydrocarbons, but not with linseed oil or petroleum.

Diethylene glycol monoethyl ether
$C_2H_5O . CH_2 . CH_2 . O . CH_2 . CH_2 . OH$

Known more correctly as 2-(β-ethoxyethoxy) ethanol it is called commercially 'carbitol' or 'dioxitol' [20]. It is made [12] by the action of ethylene oxide on the sodium compound of ethylene glycol monoethyl ether (q.v.). It is a good solvent for cellulose nitrate, shellac, kauri, mastic, sandarac, copal, elemi, coumarone, colophony, polyvinyl acetate, zein, and dyes. It partly dissolves ester gum, zanzibar and dammar, but does not dissolve cellulose acetate, polyvinyl chloride or chloroacetate, rubber or castor oil. sp. gr. 0.9996; b. p. 202° C; n_D about 1.43; flash p. 210° F (99° C); m. p. −76° C; v. p. 0.13; sp. ht. 0.55; lat. ht. 96; visc. 38 at 25° C; s. ten. 35; elec. cond. 2.5×10^{-8}; dil. rat. toluene 1.65. Miscible with water in all proportions, but not with mineral spirit. Sweet smelling, hygroscopic liquid.

It is used as a mutual solvent for coupling immiscible liquids, in cutting oils, insecticides and dry-cleaning soaps. Its good solvency for dyestuffs renders it valuable in printing inks and wood stains where its non-grain raising properties can be advantageous. It is also employed in castor oil and synthetic-based hydraulic fluids where its low temperature viscosity assists in conforming to specification needs (e.g. in aircraft brake fluids).

It is used in the manufacture of textile soaps and in dye printing.

The mono-acetate carbitol acetate is softener-solvent for cellulose nitrate and acetate. sp. gr. 1.011; b. p. 217.7°C; v. p. 0.1; flash p. 110° F; dil. rat. toluene 2.2.

It dissolves cellulose acetate, cellulose nitrate, kauri, colophony, dammar, ester gum, gum camphor, polyvinyl acetate, and chloro-acetate, low viscosity silicones, and partly cuomarone, mastic, elemi. Non-solvent for polyvinyl chloride, sandarac, copals, shellac. It is miscible with water and aromatic hydrocarbons, but not with petroleum hydrocarbons or linseed oil.

The methyl ether, known as methyl carbitol has b. p. 194.2; v. p. 0.2; flash p. 200° F (93° C)and is miscible with water in all proportions.

The higher mono-ethers of the series have the following characteristics:

Ether	b. p. (° C)	sp. gr.	n_D at 25° C
n-Propyl	150	0.911	1.413
Isopropyl	144	0.914	1.408
n-Butyl	171	0.919	1.418
Isobutyl	159	0.913	1.414
Isoamyl	181	0.900	1.420

Triethylene glycol monoethyl ether

$C_2H_5O[CH_2 . CH_2 . CH_2 . O]_2 . CH_2 . CH_2 . CH_2 . OH$

Known commercially as 'Trioxitol' or 'Ethoxytriglycol', it is a mild-smelling liquid of low hygroscopicity and is nearly the highest boiling glycol of this range. Since it possesses both ether linkage and an alcohol group it exhibits many properties common to both alcohols and ethers.

Characteristics: sp. gr. 1.030_{20}; b. p. 255.4°C; m. p. −18.7° C; flash p. (Cleveland open cup) 135° C; n_D 1.4390; max. acidity, as acetic acid, 0.02% by weight.

The commercial product is virtually free from both ethylene and diethylene glycol and it is similar in its applications to diethylene glycol monoethyl ether. It is also useful as an intermediate in plasticiser manufacture.

Toxicity is low and normal precautions only are necessary.

Diethylene glycol mono-n-butyl ether

$C_4H_9O . CH_2 . CH_2 . O . CH_2 . CH_2 . OH$

2-(2-butoxyethoxy) ethanol, known commercially as 'Butyl carbitol' [2] or 'Butyl Dioxitol' [20] is a liquid of mild odour, which dissolves cellulose nitrate, polyvinyl acetate and chloroacetate, kauri, sandarac, mastic, elemi, coumarone, colophony, but not cellulose acetate or olyvinyl chloride. It partly dissolves ester gum, dammar, zanzibar and shellac. It is without action on raw rubber. It is miscible with water, linseed oil, and hydrocarbons. sp. gr. 0.969_{20}; flash p. $172°$ F ($78°$ C); b. p. $231°$ C; n_D 1.43; v. p. 0.06; sp. ht. 0.546; lat. ht. 62; visc. 5.25; s. ten. 34; cub. exp. 0.00087; auto-ignition temp. $228°$ C.

The acetate 'Butyl Carbitol Acetate' or Butyl 'oxitre' Acetate has sp. gr. 0.981_{20}; b. p. $246°$ C; flash p. $240°$ F ($115.5°$ C); v. p. 0.04; dil. rat. xylene 1.8; water dissolves 6.5%, and it dissolves 3.7% of water at $20°$ C.

It is a solvent for cellulose acetate, colophony, ester gum, polyvinyl acetate and chloroacetate and a partial solvent for kauri, and dammar. It does not dissolve cellulose acetate, polyvinyl chloride, shellac or raw rubber. It is miscible with all the usual solvents and diluents.

Ethylene glycol monoisopropyl ether
$(CH_3)_2 . CH . O . CH_2 . CH_2 . OH$

Known also as 2-isopropoxy ethanol or isopropyl oxitol [20] this solvent is a clear, mobile, agreeable smelling liquid exhibiting both ether and alcohol characteristics.

sp. gr. $0.908_{20\ 20}$; b. p. $142.8°$ C; flash p. (Pensky Martens) $130°$ C; n_D 1.1410; dil. rat. toluene 4.4. If forms an azeotrope with water containing 60% water 40% solvent, this possessing b. p. $99.4°$ C. It is miscible in all proportions with water, aromatic and aliphatic hydrocarbons, esters, ketones, and alcohols and is useful as a coupling agent with immiscible liquids. It is used as a dye solvent in both surface coatings and textile dyeing or printing and it is also valuable as a high boiling solvent in cellulose nitrate, ethyl cellulose, benzyl cellulose and cellulose acetate-butyrate lacquers. Generally, it may be regarded as an economic alternative to ethylene glycol monoethyl ether with a slightly higher boiling point and slower evaporation rate.

It is considered to possess similar toxic and dermatitic properties to other glycol ethers in the range.

Dioxane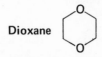

Dioxane, or 1,4 diethylene dioxide, was originally produced as an unwanted by-product in the process for manufacturing the ethers of ethylene glycol and of diethylene glycol. It results by the loss of two molecules of water from two molecules of glycol, and is a di-ether. It is made by polymerising ethylene oxide with caustic alkali or by distilling ethylene glycol with phosphoric acid, sulphuric acid, zinc chloride or benzene sulphonic acid; the process is made continuous by adding fresh glycol to the boiling mixture to replace that which is used up [15].

Physical Characteristics: sp. gr. 1.0355_{20}; m. p. $11.8°$ C; b. p. $101.3°$ C; v. p. 29; lat. ht. 86; sp. ht. 0.41; n_D 1.423; flash p. $52°$ F ($11°$ C); s. ten. 37 at $25°$ C; cub. exp. 0.00103; elec. cond. below 2×10^{-8}; it is miscible with water in all proportions. It forms an azeotrope with 18% of water: b. p. $87.8°$ C. A commercial product is available having: b. r. $94°-110°$ C; sp. gr. 1.030_{20}; flash p. $41°$ F ($5°$ C); visc. 1.2 at $25°$ C; dielec. const. 2.2; s. ten. 37 at $25°$ C. A colourless liquid with an odour of butyl alcohol. It is dangerously toxic. Dioxane is on the list of poisonous substances coming within the schedule of industrial diseases under the Workmen's Compensation Act, Home Office Statutory Rules and Orders, 1934, No. 551. Miscible with water, petroleum and aromatic hydrocarbons, castor and linseed oils, and most organic solvents. Stable to light but forms an explosive peroxide in air.

The following constant-boiling mixtures are known:

Dioxane (%)	(%)	b.p. ($°$ C)
80	water 20	87.0
40	m-xylene 60	136.0
2.5	Isoamyl-alcohol 97.5	131.3

Dioxane dissolves ester gum, polystyrene, glyceryl phthalate, polyvinyl acetate, chloride and chloroacetate, shellac, coumarone, manila, dewaxed dammar, pontianac, elemi, kauri, sandarac, mastic, guaiac resins [15] rubber, gutta percha resin, plextol, lignin, cellulose acetate with an acetyl content of 36% or more, cellulose aceto butyrate, ethyl and benzyl cellulose, and, in the presence of alcohol, cellulose nitrate containing 10.8 to 13.8% of nitrogen; water increases the solvent properties of dioxane for both cellulose acetate and nitrate.

References
[1] *Cf.* D.R.P. 404999 (1919).
[2] Union Carbide, London.
[3] *Paint, Oil and Chem. Review*, April, 1928, p. 11.
[4] Brit. Pat. 286850. *Cf.* Brit. Pat. 365589.
[5] U.S. Pat. 1614883.

[6] Brit. Pat. 271169; F.P. 610282.
[7] C.P. 265191 (1926); B.P. 322037; B.P. 354357; B.P. 367353.
[8] Brit. Pat. 236379; D.R.P. 403643 (1921). *Cf.* B.P. 374864.
[9] Brit. Pat. 402438; U.S.P. 1954336
[10] D.R.P. 403643; Brit. Pat. 236379.
[11] *Ind. Eng. Chem.*, 1927, p. 474.
[12] U.S. Pat. 1633927.
[13] Brit. Pat. 558776.
[14] U.S. Pat. 1681861 (1928).
[15] Brit. Pat. 275653.
[16] Brit. Pat. 311671; 328083.
[17] (1) Spangler and Davies. *Ind. Eng. Chem.,* anal., 1943, Feb., p. 96.
     (2) American Bureau of Standards. *Ind. Eng. Chem. News.*, 1926, No. 4, p.
     1.
[18] U.S. Pat. 2,272,269–70.
[19] I.C.I., Brit. Pat. 560,770.
[20] Shell International Chemical Co. Ltd.

CHAPTER SIX

Cyclohexane derivatives

Cyclohexanol.

Cyclohexanol, or hexahydrophenol, is a product obtained by the catalytic hydrogenation of phenol, and consists of a mixture of two geometric isomers. Phenol, completely freed from sulphur compounds, is treated with hydrogen gas under at least four atmospheres pressure, in the presence of a nickel catalyst, at temperatures ranging between 160°–200° C [1]. The reaction takes place in two stages, tetrahydrophenol first being formed. If the temperature be allowed to rise too high relatively to the pressure, dehydrogenation of the hydroxyl group takes place and cyclohexanone is formed; this fact forms the basis for the manufacture of cyclohexanone.

Cyclohexanol is an oily liquid, having a persistent camphoraceous, amyl alcohol odour, strong, but not highly objectionable.

Physical Characteristics: sp. gr. 0.947_{20}; n_D 1.465–1.466; flash p. 155° F (68° C); m. p. (pure) 25°C, (tech.) 18° C min; b. p. 160° C; b. r. 158° –163° C; solubility in water 6% at 20° C; dissolves 12% of water at 20° C; neutral and non-phenolic; visc. 4.6 at 25° C; dielec. const. 15 at 25° C; lat. ht. 108; sp. ht. 0.417; evap. rate 400; hydroxyl no. 520/560.

Pure cyclohexanol is not a solvent for either cellulose acetate or nitrate; technical preparations containing cyclohexanone dissolve some forms of the acetate as well as the nitrate.

It is a solvent for cellulose ethers, ester gum, glyceryl phthalate resin, benzyl abietate, mastic, shellac, coumarone, colophony, kauri, manila, polyvinyl chlorocyanide, low viscosity silicones, bakelite A, gutta percha resin, zein, and for basic dyes. Its solvent properties for rubber, copal, dammar and elemi are poor. It finds use in amine cured epoxy finishes where it assists in reducing such faults as cratering and pinholing; both air drying and stoving epoxy finishes can be improved by additions but care must be used in selecting the percentage. It is miscible with linseed oil, petroleums and aromatic hydrocarbons. It forms the following azeotropic mixtures.

Cyclohexanol (%)	(%)	b. p. (°C)
45	furfural 55	156
41	camphene 59	152
35	pinene 65	150
36	pentachloroethane 64	158
71	cymene 29	159

Cyclohexanyl acetate

O.OC.CH₃

Cyclohexanyl acetate is the acetic ester of cyclohexanol, made by direct esterification. The quality of technical preparations varies considerably, and depends largely on the degree of esterification attained; good technical quality answers to the following specification: sp. gr. about 0.96; b. r. 170°–177° C; flash p. 139° F (59.5° C); n_D 1.438–1.440. esters 80% min. evap. range 77. visc. 2.3 at 25° C. dil. rat. 2.5 toluene, 3.0 xylene, 1.5 white spirit. it dissolves 1.4% of water at 20° C.

Cyclohexanyl acetate has a strong, stale, fruity odour resembling amyl acetate; it is non-toxic. It is a powerful solvent for cellulose nitrate and cellulose ethers, giving viscous solutions; also for ester gum, colophony, dammar, elemi, manila, sandarac, kauri, congo, basic dyes, blown oils, plexol, raw rubber, rubber chloride, gutta percha resin, urea-formaldehyde resin, polystyrene, glyptals, carnauba, metal resinates, driers and albertols.

Methylcyclohexanol

OH
CH₃

Methylcyclohexanol, hexahydromethylphenol, or hexahydrocresol is a mixture of three isomeric secondary alcohols, each of which can exist in two geometric modifications, the predominating constituent being one of the ortho-members as indicated in the formula.

It is prepared by the hydrogenation of cresol in a manner similar to that for the manufacture of cyclohexanol, but technical products vary according to the relative proportions of ortho-, meta-, and para-cresols present in the cresol hydrogenated. Methy cyclohexanol is similar to, but not as powerful as, cyclohexanol and it finds uses in amine cured epoxy finishes in a similar manner to cyclohexanol.

The following are the physical characteristics of a good technical quality containing the six isomers together with traces of cyclohexanol: sp. gr. 0.925; n_D 1.455–1.465; flash p. 155° F (68° C); s. p. −50° C; b. r. 173°–175° C; visc. 28 at 25° C; evap. range 807; R.W. phenol coefficient 1.5; solubility in water 3%; hydroxyl no. approx 450;

British Standard Specification, BS 2714:1956 for methylcyclohexanol requires: sp. gr. 0.920–0.932, $(0.916–0.928)_{20}$; b. r. 165°–180° C 95% min; residue 0.02% max; acidity 0.01% max. as acetic; special test for phenol; hydroxyl content 14–15%.

Methylcyclohexanol is an oily liquid having a camphoraceous odour similar to that of cyclohexanol.

As a solvent it is similar to, but not so powerful as, cyclohexanol.

The following figures have been given for the pure cis-transisomers[5]:

	n^{29}	sp. gr. $_{30}$ 4
cis-2-Methyl Cyclohexanol	1.4694	0.9274
trans-2-Methyl Cyclohexanol	1.4616	0.9174
cis-4-Methyl Cyclohexanol	1.4614	0.9173
trans-4-Methyl Cyclohexanol	1.4561	0.9040

Methylcyclohexanyl Acetate

Technical methylcyclohexanyl acetate consists of the mixture of esters obtained by the acetylation of technical methylcyclohexanol. Technical material has approximately the following characteristics: esters 75–80%; sp. gr. 0.94–0.95; b. r. 172°–192° C; flash p. 143° F (62° C); visc. 2.36 at 25° C; evap. range. 5 (butyl acetate = 100); n_D 1.42–1.44; dil. rat. 2.25 toluene, 2.0 benzene, 1.5 white spirit.

Its solvent properties are much the same as those of cyclohexanyl acetate, but it is somewhat slower in action; the solutions are usually more viscous and take longer to dry. Methylcyclohexanyl acetate is a good solvent for cellulose nitrate, also for colophony, ester gum, manila, mastic, benzyl abietate, elemi, dammar, kauri, bakelite, basic dyes, raw rubber, tung oil, waxes and bitumen. It has some solvent action on cellulose acetate, shellac and hard copals.

Cyclohexanone

Obtained by the catalytic hydrogenation of phenol at high temperatures, also by the dehydrogenation of cyclohexanol, by oxidation of this with bichromate mixture and by treating a mixture of cyclohexanol and phenol with a hydrogenation catalyst [2]. It is a colourless liquid, having an odour suggestive of peppermint and acetone.

Physical Characteristics: Technical material [3]: 85% min. ketone. sp. gr. 0.950; b. r. $150°-158°$ C, 95%; $n_D1.443-1.451$; flash p. $117°$ F ($47°$ C); visc. 2.2 at $25°$ C; evap. range 40; dil. rat. 6.5 benzene, 7.25 toluene and xylene, 1.25 white spirit; chief impurity, cyclohexanol. Pure cyclohexanone has b. p. $156.5°$ C; m. p. $-45°$ C; sp. ht. 0.43; n_D 1.450; elect. cond. $5-8 \times 10^{-8}$; dielec. const. 18.2 at $25°$ C; water dissolves 5 at $20°$ C; it dissolves 8.7% of water at $20°$ C.

British Standard Specification, BS 2711:1967 for methylcyclohexanone requires: sp. gr. 0.949–0.951 ($0.946-0.948_{20}$, $0.943-0.945_{25}$); b. r. 95% between $153°$ C and $157°$ C; flash p. min. $105°$ F ($41°$ C approx); max. ash 0.01%; not alkaline; max. acidity 0.01%; ketones 98% min.

It is a good solvent for cellulose nitrate, yielding solutions of approximately the same viscosity as those of ethyl lactate and less viscous than those of ethylene glycol monoethyl ether. Its solutions of cellulose nitrate are exceptionally tolerant to dilution with toluene.

Cyclohexanone is a solvent for cellulose acetate and ethers, colophony, ester gum, kauri, elemi, manila, shellac, raw rubber, rubber chloride, polyvinyl chloride, acetate and chloroacetate, polystyrene, bakelite, cyclohexanone-formaldehyde, rezyl, glyptal and coumarone resins, basic dyes, blown oils, bitumen, waxes and albertols. It partly dissolves congo, dammar, and it is miscible with most organic solvents. It dissolves 37% w/v of gamma benzene hexachloride at $20°$ C. It attacks polythene and Perspex. It forms the following azeotropic mixtures:

Cyclohexanone (%)	(%)	b. p. ($°$ C)
55	tetrachloroethane 45	159
39	trichlorhydrine 61	160
56	camphene 44	151

Methycyclohexanone

Methycyclohexanone is a colourless liquid having an odour of peppermint; it is prepared, similarly to cyclohexanone, from the cresols and has similar solvent properties. The technical product [3], which consists of a mixture of the three isomers, ortho-, meta-, and para-, has the following approximate characteristics: sp. gr. 0.92; n_D 1.44–1.45; flash p. 130° F (55° C); b. r. 160–175° C; evap. range 47; visc. 1.7 at 25° C; dil. rat. 5.5, benzene, 5.5, toluene, 7.0 xylene, 1.25 white spirit; ketones 85% min, water dissolves 3%. For some uses a more highly refined material is commercially available [3].

A product having a ketone-content of 100% has b. r. 164°–172° C; sp. gr. 0.925; visc. 1.8. the pure ketone has b. p. 170° C; m. p. −70° C.

Isophorone

Isophorone or 3,5,5,-trimethylcyclohex-2-ene-1-one is now available technically [4]. It is a high boiling, stable, colourless, mild odoured ketone and is a good solvent for cellulose nitrate and some vinyl resins. It has sp. gr. 0.923_{20}; b. p. 215° C; m. p. −8° C; flash p. 205° F (96° C); v. p. 0.25; dil. rat. 5.7 toluene, 5.1 xylene, 0.7 petroleum. dissolves 4.0% of water; water dissolves 1.2%.

References

[1] *Cf.* Hilditch. "Catalytic Processes," p. 295. Chapman & Hall, 1937.
[2] Brit. Pat. 310055.
[3] Howards and Son, Ilford, London. Trade Mark names, 'Sextone,' 'Sextone B,' 'Sextone Z,' 'Sextate,' 'Sextol.'
[4] British Industrial Solvents, London, W.1.
[5] *J. C. S.*, 1949, p. 1718.
[6] British Standards Institution, 2 Park Street, London, W.1.

CHAPTER SEVEN

Chloro-compounds

Ethylene Chlorhydrin $Cl.CH_2.CH_2OH$

Ethylene chlorhydrin, known also as glycol chlorhydrin and as 2-chloroethyl alcohol, is a colourless liquid of mild odour.

This substance is an 'intermediate' involved in the manufacture of the glycol ether group of solvents and is produced from ethylene derived from producer gas, from ethyl alcohol, or from the gases arising in the cracking of petroleum. The ethylene is converted into ethylene chlorhydrin by treatment with chlorine water [1]. Water is circulated through a saturating tower wherein it is treated with chlorine gas, thus forming a solution of hypochlorous and hydrochloric acids. It then passes to a second tower and comes into contact with the ethylene which, at temperatures ranging from $0°$ to $20°$ C, combines with the hypochlorous acid to form ethylene chlorhydrin, whilst the dilute hydrochloric acid solution, which remains, is continually returned to the chlorinating tower for further treatment with chlorine until its content of hydrogen chloride reaches about 3½%; it is then removed. From the 8 to 10% solution thus obtained the ethylene chlorhydrin is isolated by fractional distillation.

It is also prepared in a somewhat similar fashion [2] by agitating chlorine and ethylene with water until the hydrochloric acid content reaches about 15%, whereafter the solution is neutralised with lime and the ethylene chlorhydrin distilled out. Some difficulty is presented in isolating the chlorhydrin from the reaction mixture, one method consists in first extracting the by-product, ethylene dichloride with petroleum spirit and then the chlorhydrin with a mixture of isopropyl ether and isopropyl alcohol. Another method [4] is that of neutralising the hydrochloric acid with ammonia and then concentrating the chlorhydrin by distillation.

Ethylene chlorhydrin forms a constant-boiling mixture with water, boiling at $97.8°$ C and containing 42% of the chlorhydrin by weight. The anhydrous material (96–98%) has the following characteristics: sp. gr. 1.20–1.21 at $20°$ C; b. r. $125°$–$132°$ C; flash p. $131°$ F ($55°$ C); b. p. 128.6; lat. ht. 123; n_D 1.444 at $15°$ C; m. p. $-69°$ C; v. p. 5; dielec. const. 25.8 at $25°$ C; It is clearly miscible with benzene, water, alcohol, etc. The chief impurity is ethylene dichloride.

Ethylene chlorhydrin is a solvent for cellulose acetate, and the solutions have a high tolerance for water. Cellulose acetates containing over 50% of combined acetic acid will dissolve in mixture of equal parts of ethylene chlorhydrin and water. It is a solvent for shellac.

Formerly ethylene chlorhydrin was regarded as harmless but observations and experiments have shown it to be a very dangerous material possessing unsuspected toxic properties. Poisoning can occur by inhalation and by absorption through the skin. It has been found to be a metabolic poison with a specific effect on the nervous system; it appears also to affect the kidneys and may cause asphyxia [5].

Glyceryl Monochlorhydrin. $Cl . CH_2 . CH(OH) . CH_2 . OH$

The commercial article is a mixture of two isomers of which the preponderating one is represented by the formula given above. Monochlorhydrin is not largely used in industry, as its solvent powers are not very great; it is hygroscopic and develops acidity on prolonged contact with moisture; it also darkens with age. sp. gr. 1.28–1.35; b. r. 213°–228° C (decomposes).

It is miscible with water and most organic solvents, but not with hydrocarbons or vegetable oils.

Monochlorhydrin is a solvent for cellulose acetate and is useful for preparing aqueous solutions of cellulose acetate having up to 60% of water. It dissolves glyceryl phthalate resins and has a limited solvent action on nylon, ester gum, benzyl abietate, shellac and mastic. It does not dissolve cellulose nitrate, copals, copal ester, cumarone and hard resins.

Monochlorhydrin is prepared by the partial esterification of glycerine with hydrochloric acid.

Glyceryl Dichlorhydrin $Cl . CH_2 . CH(OH) . CH_2 . Cl$

Technical glyceryl dichlorhydrin is a mixture of 1,3-dichloropropan-2-ol and 1,2-dichloropropan-3-ol, the former (shown above) preponderating. It is a colourless liquid, having a weak odour of chloroform, and is not readily inflammable.

Like other chlorhydrins, it tends to liberate hydrochloric acid in the presence of moisture and also to turn brown with age. It possesses powerful solvent properties, but its defects severely restrict its use.

Physical Characteristics: sp. gr. 1.34–1.38; b. r. 95% between 174° and 176° C; b. p. (1.3) 174° C, (1.2) 183° C; n_D 1.47–1.48; flash p. 165 F (74° C); v. p. 7; dielec. const. 12; solubility in water about 10%.

It is miscible with castor and linseed oils, aromatic hydrocarbons and most organic solvents, but not with petroleums or turpentine. It is a solvent for cellulose acetate, ethyl cellulose and for some forms of cellulose nitrate, for ester gum, copal ester, glyceryl phthalate resin, benzyl abietate, shellac, coumarone, mastic, elemi, dammar, manila, kauri, soft copals and caoutchouc. It attacks nylon.

It forms a constant-boiling mixture with 50% of ethyl lactate boiling at 143° C, and with 61% of cyclohexanone boiling at 160° C.

Epichlorhydrin $CH_2-CH-CH_2.Cl$
                          $\diagdown O \diagup$

Epichlorhydrin is a highly mobile, colourless liquid having an odour weaker than, but resembling, that of chloroform; it is narcotic but not poisonous. Although it is an excellent solvent for lacquer work, it possesses the unfortunate property of liberating hydrochloric acid in contact with moisture.

It is prepared from the dichlorhydrins by the action of alkalis; it is non-miscible with water and with petroleum hydrocarbons, but miscible with the aromatic hydrocarbons and all the usual solvents. sp. gr. 1.19–1.20; b. p. 116° C; m. p. −48° C; n_D1.438; lat. ht. 0.83; flash p. 105° F (40.5° C); visc. 1 at 25° C; dielec. const. 23; elec. cond. 5.4×10^{-8}; v. p. 13.

It is a powerful solvent for cellulose nitrate and acetate, ester gum, glyceryl phthalate resin, benzyl abietate, coumarone and soft resins; a moderately good solvent for ethyl cellulose, shellac, mastic, kauri and manila; it does not dissolve hard copals or copal ester. It forms the following azeotropic mixtures:

Epichlorhydrin (%)	(%)	b. p. (° C)
23	n-Propyl alcohol 77	96
40	Isobutyl alcohol 60	105
81	Isoamyl alcohol 19	115
26	Toluene 74	108
52	Tetrachloroethylene 48	110

Methylene Dichloride CH_2Cl_2

Methylene dichloride or dichloromethane is a highly volatile solvent which is produced, along with chloroform (q.v.), carbon tetrachloride and methyl chloride by the chlorination of methane [6]. It is a solvent for cellulose esters and has been used to raise the flash-point of lacquers [32], but this use is not recommended as, like most chlorinated paraffins, methylene dichloride is narcotic.

Characteristics: sp. gr. $1.335_{15\,4}$, $1.328_{15\,4}$; b. p. $39.8°$ C; sp. ht. p.288; lat. ht. 78; v. p. 145_0, 230_{10}, 355_{20}, 530_{30}; visc. 0.463_{15}; n_D 1.424; s. p. $-96.7°$ C; evap. range 0.71; s. ten. 28; therm; cond. 0.092; cub. exp. 0.00137; dissolves water $0.09_0\%$, 0.12_{10}, 0.17_{20}; water dissolves 2.363_05, 2.122_{10}, 2.000_{20}, 1.969_{30}; elec. cond. 4.3×10^{-11}.

British Standard Specification, BS 1994:1953 for methylene dichloride requires: sp. gr. 1.328–1.338 $(1.321–1.331)_{20}$; b. r. $39.0°$–$40.5°$ C 95%; residue 0.010% max. Water 0.050% max; acidity 0.001% max. as HCl; free from free chlorine; colourless.

It is a solvent for polyvinyl acetate, chloride and chloroacetate, methyl and ethyl cellulose, rubber, bitumen, pitch, caoutchouc, oils, waxes, and other resins. Its solvent power is often increased by adding an alcohol such as methanol.

It frequently contains traces of methyl chloride, which lowers the boiling-point.

It forms an azeotrope with 1.8% of water, b. p. $38.1°$ C.

Chloroform $CHCl_3$

Chloroform, being of a highly narcotic nature, does not find wide application as a lacquer solvent, but is possesses very high solvent powers and has been used for special purposes such as the manufacture of art silk.

It can be made by a variety of methods, and the time-honoured one of distilling alcohol or acetone with bleaching powder is still used; a variation consists in chlorinating anhydrous alcohol with dry chlorine gas and treating the chloral hydrate thus formed with alkali. [7] It can also be made by treating acetaldehyde with bleaching powder solution. [8]. A process now attaining considerable importance is the direct catalytic chlorination of methane. [9] Mixtures of methane, methyl chloride and chlorine are passed through earthenware reaction chambers at temperatures ranging from $350°$–$650°$ C, whereby mixtures of chloroform, carbon tetrachloride and methylene dichloride are obtained.

Chloroform has the following characteristics: sp. gr. 1.4892_{20}, 1.4798_{25}; $n_D 1.4458$; b. p. $61.2°$ C; m. p. $-63°$ C; v. p. 162; lat. ht. 59; sp. ht. 0.225; visc. 0.556_{20}, 0.538_{25}; s. ten. 27.2_{20}, 26.55_{25}; elec. cond. $< 10^{-10}$; dielec. const. 4.64; therm. cond. 0.075; cub. exp. 0.0014; evap. range 0.56; non-inflammable. solubility in water 0.8%, dissolves 0.097% of water.

Chloroform slowly becomes acid in the presence of light and moisture: a small quantity of alcohol is usually added to prevent this; the British Pharmacopoeia specifies 2%. It dissolves amylose triacetate, cellulose acetate and propionate, benzoate, dinaphthenate and ethyl cellulose, also rubber, gutta percha, methyl methacrylate and most resins, but not shellac. It dissolves 24% w/v of gamma benzene hexachloride.

The following azeotropic mixtures are known:

Chloroform (%)	(%)	b. p. ($^\circ$ C)
87.5	Methyl alcohol 12.5	53.5
93	Ethyl alcohol 7.0	59.4
72	n-Hexane 28	60
90	Acetone 20	64.7
22	Methyl acetate 78	64.5

Carbon Tetrachloride CCl_4

Carbon tetrachloride or tetrachloromethane is one of the most widely used solvents of this group. It is a colourless, non-inflammable liquid, having an odour somewhat similar to that of chloroform; it is dangerously toxic and generates poisonous gases, namely, phosgene, hydrochloric acid, and chlorine, when in contact with burning substances, or with water at high temperatures. It reacts violently with burning sodium and may react explosively with aluminium powder [37].

It is made from carbon bisulphide by the action of chlorine in the presence of sulphur or iodine, [10] or of sulphur chlorides [11] in iron vessels the use of which prevents the formation of the highly toxic trichloromethylsulphur chloride; also by the chlorination of methane or methylene chloride, [12], acetylene [34] and perchlorethylene, [35] and by passing chlorine through incandescent carbon or coke. [13]

British Standard Specification, BS 575:1966 for carbon tetrachloride requires: sp. gr. 1.601–1.607 (1.594–1.600$_{20}$, 1.586–1.592$_{25}$); b. r. 76.0–77.5° C; max. residue 0.01%; pH of aqueous extract 6.0–7.5; no free chlorine; max. sulphur compounds 0.05%.

Pure carbon tetrachloride has sp. gr. 1.604, 1.595$_{20}$; b. p. 76.75° C; m. p. -23° C; v. p. 90; n_D 1.460; evap. range 0.33; visc. 0.968$_{20}$, 0.902$_{25}$; s. ten. 26.75$_{20}$, 26.15$_{25}$; elec. cond. 4×10^{-18}; dielec. const. 2.2; dielec. strength 33,000 volts min; therm. cond. 0.068; cub. exp.

0.0012; lat. ht. fusion 600 cal/g. mol; water dissolves $0.08\%_{20}$. dissolves $0.008\%_{20}$ of water.

It tends slowly to liberate hydrochloric acid on contact with water, but has no action, when dry, on iron or nickel; it attacks copper and lead slightly.

It dissolves some varieties of cellulose acetate and dinaphthenate, but not the nitrate or propionate; it is a good solvent for ethyl cellulose, coumarone, benzyl resin, mastic, polystyrene, dammar, rosin, ester gum, elemi, albertols, bitumen, thiourea resins, rubber and rubber heptachloride, and partly dissolves sandarac, soft copals, kauri. It is a moderate solvent for DDT and gammexane (7%). It does not dissolve shellac, accroides, hard copal, polyvinyl chloride and chloracetate.

The following azeotropic mixtures are known:

Carbon tetrachloride (%)	(%)	b. p. ($^{\circ}$C)
57	Ethyl acetate 43	74.8
84	Ethyl alcohol 16	64.9
79	Methyl alcohol 21	55.7
89	n-Propyl alcohol 11	72.8
94.5	Isobutyl alcohol 5.5	75.8
86	Isopropyl alcohol 14	67.0
71	Methyl ethyl ketone 29	73.8

Sym-Dichlorethane $CH_2Cl.CH_2Cl$

Sym-Dichlorethane, ethylene dichloride or 1,2-dichlorethane is a colourless mobile liquid with a sweet taste and an odour similar to that of chloroform; it is toxic and produces dangerous physiological effects like those of chloroform. Air saturated at 20° C with dichlorethane contains about 10% by volume; the odour of dichlorethane is quite marked at 0.1%, at which concentration symptoms of poisoning can occur after several hours exposure.

In its manufacture by the direct chlorination of ethane, the two gases, in equal volumes, are led over activated carbon at $100^{\circ}-300^{\circ}$ C, ethyl chloride and trichlorethane are also formed in small quantities. [33] Also made from ethylene and chlorine by treatment with calcium chloride, [14] or with the chloride of iron, copper, or antimony at $30^{\circ}-120^{\circ}$ C [15], or by passing ethylene gas into liquid chlorine at low temperature. In the manufacture of ethylene glycol it is formed as an intermediate product. It is slowly hydrolysed by water, and causes slight corrosion of metals. It is one of the most stable of the chlorinated paraffins but it tends to turn yellow when stored.

Characteristics: sp. gr. 1.2527_{20}, b. p. 83.6; m. p. $-36°$ C; v. p. 62; n_D 1.4448; sp. ht. 0.308; lat. ht. 77; ht. comb. 2720; flash p. 65° F (18° C); ign. t. 450° C (burns without difficulty); evap. range 0.27; visc. 0.829_{20}, 0.775_{25}; s. ten. 32.45_{20}, 31.75_{25}; therm. cond. 0.083; cub. exp. 0.0012; dielec. const. 10; elect. cond. under 1×10^{-8}; water dissolves $0.8\%_{20}$; dissolves $0.15\%_{20}$ water; dil. rats. 8.6 acetone, 5.4 ethyl acetate, 4.3 isopropylacetate, 3.7 butyl acetate, 9.2 methyl cellosolve, 7.6 cellosolve.

The following azeotropes are known:

Dichlorethane (%)	(%)	b. p. (° C)
93	Water 7	71.5
83.5	Methyl alcohol 16.5	59.5
78.6	Ethyl alcohol 21.4	70.5
78	Ethyl alcohol 17 Water 5	66.7

It will not dissolve cellulose nitrate or acetate except after admixture with alcohols, glycol ethers or ethyl acetate. It is a solvent for rubber, and partly for soft copals, sandarac, mastic, elemi, guaiac, japan wax, lead linoleate. It readily dissolves manila, dammar, coumarone, ester gum, camphor gum, rosin, gum lac, bakelite, albertols, urea formaldehyde and glyceryl phthalate resins, polystyrene, polyvinyl acetate, chloroacetate, methyl methacrylate, raw rubber, zinc resinate, benzyl abietate, benzyl cellulose, ethyl cellulose, turpentine, paraffin wax, gilsonite, asphalt, bitumen, mineral oils, cottonseed oil, castor oil, linseed oil. It does not dissolve congo, kauri, pontianac, shellac, beeswax, carnauba wax or nylon. It attacks rayon, polyvinyl chloride and Perspex.

1,1,2-Trichloroethane CH_2Cl . $CH Cl_2$

Known also as β-trichloroethane, this toxic solvent should not be confused with the much safer 1,1,1 isomer. The TLV is 10 ppm and it is therefore more toxic than many of the common chlorinated solvents, notable exceptions being tetrachloroethane (5 ppm) and carbon tetrachloride (10 ppm). It is narcotic with an odour resembling that of chloroform. Trichloroethane is produced by chlorinating dichlorethane at 50° C in the presence of ethylene, or an iron-aluminium catalyst and ultra-violet light. It is a good solvent for cellulose acetate, raw rubber, some types of synthetic rubber and most oils, fats and waxes.

Sp. gr. 1.443 at $20°$ C; b. p. $113.7°$ C; v. p. 16.7; n_D 1.47; s. p. $-35°$ C; lat. ht. 68; sp. ht. 0.266; visc. 1.2; evap. range 0.12; therm. cond. 0.0778; cub. exp. 0.001; water dissolves 0.45% and it dissolves 0.03% of water at $20°$ C.

1,1,1-Trichloroethane $CH_3 CCl_3$

1,1,-trichloroethane, or methyl chloroform, is a colourless non-inflammable liquid having a sweet odour not unlike carbon tetra-chloride. It has a mild narcotic effect, lower than the majority of common chlorinated solvents and in addition has been shown to be free from chronic side effects [40]. The TLV is 350 ppm. It has therefore found wide acceptance as a safer replacement for both non-inflammable and inflammable solvents in cold cleaning and in formulation. Suitably stabilised grades of 1,1,1-trichloroethane can also be used as safer alternatives to traditional vapour cleaning solvents such as trichloro-ethylene and perchloroethylene.

1,1,1-trichloroethane is manufactured from 1,1-dichloroethylene by hydrochlorination. The pure solvent is stable to light (as is the stabilised product) but tends slowly to liberate hydrochloric acid on contact with water. It has no action when dry on iron, copper or nickel, but reacts readily with aluminium or its alloys. Commercially available solvents such as 'Genklene' contain stabilisers to inhibit this reaction. These have an effect on the physical properties of the solvent.

	Unstabilised Solvent	Stabilised Solvent
sp. gr. at $25°$ C	1.332	1.300–1.327
n_D at $20°$ C	1.4377	1.4315–1.4335
b. p. (Range)	$74.1°$ C	$71°$–$81°$ C

The pure material has sp. gr. $1.3366_{25\ 4}$; $n_D 1.4377$; b. p. $74.1°$ C; m. p. $-32.7°$ C; sp. ht. 0.280; lat. ht. 54.2; v. p. 38_0, 64_{10}, 104_{20}, 106_{30}; cub. exp. 0.00129; dielec. const. 9.3; therm. cond. 0.08; K.B. value 124.

It is a solvent for rubber, chlorinated rubber, bitumen, mastic, mineral and vegetable oils, waxes and greases, stearic acid, lanolin, silicone oils, polyvinyl acetate, methyl methacrylate monomer, polystyrene and acrylic resins. It will not dissolve cellulose nitrate, cellulose acetate, rosin, shellac, nylon, polypropylene or polyethylene. PVC will not dissolve but the solvent will extract plasticisers from flexible PVC products. It is a useful solvent for adhesives in its own right or mixed with methylene chloride and in addition can be used to increase the flash point of many inflammable solvents.

Sym-Tetrachlorethane CHCl$_2$. CHCl$_2$

Formerly called acetylene tetrachloride, and known under the name of 'Westron,' [23] this solvent is a colourless mobile liquid having an odour resembling those of carbon tetrachloride and chloroform. It is the most powerful solvent of its class.

Tetrachlorethane, like most of the substances of this class, is dangerously toxic [16] and must only be employed in enclosed apparatus or in well-ventilated spaces, its use in lacquers and paints is in consequence not advisable. It causes jaundice and enlargement of the liver, the diseases being notifiable under the Factory and Workshop Act, section 73, Order No. 1170 of November, 1915.

It attacks iron, copper, nickel, and lead only slightly but aluminium vigorously. In the presence of moisture it slowly liberates hydrochloric acid.

sp. gr. 1.600–1.602; n_D1.494; b. p. 145.5° C; v. p. 5; sp. ht. 0.257; lat. ht. 55; s. p. −42.5° C; therm. cond. 0.078; cub. exp. 0.001; evap. range 0.03; visc. 1.7° C; dielec. const. 7.8; water dissolves 0.3% and it dissolves 0.03% of water at 25° C; elec. cond. 4×10^{-7}.

Tetrachlorethane is a solvent for some forms of cellulose acetate and for acetobutyrate also for colophony, ester gum, elemi, dammar, mastic, rubber, rubber chloride, bitumen, pitch, canada balsam, lanoline, paraffin wax, seekay wax, stearic acid, fluid silicones and petroleum jelly but not for cellulose nitrate, shellac, tragacanth, copals or sandarac. It dissolves about 1% of sulphur.

It is prepared by the interaction of acetylene and chlorine, but the direct reaction is dangerously explosive. In order to moderate the reaction, it is sometimes conducted in a cooled bed of sand or by passing alternately acetylene at 60°–80° C and chlorine at 80° −100° C into antimony pentachloride, [17] or by passing the gases separately but simultaneously over a catalyst composed of iron turnings mixed with quartz and sprinkled continuously with tetrachlorethane. [18]. Commercial production is by reaction of Cl$_2$ with acetylene in solution in tetra using FeCl$_3$ as a catalyst.

Another system consists in circulating a dilute solution of antimony chloride in tetrachlorethane between two vessels in one of which the liquid is treated with acetylene and in the other chlorine. A double compound of acetylene and antimony chloride forms in the first vessel and reacts with the chlorine in the second vessel to form tetrachlor-ethane and reliberate the antimony chloride.

The following azeotropic mixtures are known:

Tetrachlorethane (%)	(%)	b. p. ($^{\circ}$ C)
91	Glycol 9	145
45	Cyclohexanone 55	159
68	Isoamyl acetate 32	150

Tetrachlorethane has been known as 'Tetraline,' but this name has fallen into disuse, since it caused confusion with the name 'Tetralin' applied to tetrahydronaphthalene.

Pentachlorethane $CHCl_2$. CCl_3

This is a colourless, non-inflammable, toxic liquid prepared by the catalytic chlorination of trichlorethylene, and by the action of chlorine on sym-tetrachlorethane in actinic light. [19] sp. gr. 1.68_{20}; $n_D 1.503$; b. p. 160.5° C; v. p. at 22° C, 2.7; sp. ht. at 20° C, 0.21; lat. ht. 44; s. p. -29° C; evap. range 0.03; therm. cond. 0.075; cub. exp. 0.0009; visc. 2.2 at 25° C; dielec. const. 3.6; water dissolves 0.05% and it dissolves 0.03% of water at 25° C.

Pentachlorethane is a solvent for some forms of cellulose acetate and for colophony, ester gum, elemi, dammar, mastic, rubber, alloprene, bitumen, canada balsam, lanoline, paraffin wax, seekay wax, stearic acid, sulphur and petroleum jelly but not for cellulose nitrate, shellac, tragacanth, copals or sandarac. The following azeotropic mixture is known: Pentachlorethane 85%, glycol 15%, b. p. 154.5° C.

Sym-Dichlorethylene $CHCl = CHCl$

Sym-Dichlorethylene, sometimes termed acetylene dichloride, is not to be confused with ethylene dichloride, which is sym-dichlorethane. It is a colourless liquid, toxic, but apparently not dangerously so. The cold liquid is non-inflammable, but the hot vapour can be ignited, it burns with a cold flame which extinguishes itself; in practice there is no danger from fire. The commercial product is a mixture of two stereoisomers, the trans isomer, present to the extent of about 40%, boils at 48.3°, and the cis at 59.8° C.

Dichlorethylene is manufactured by the partial and carefully regulated chlorination of acetylene, a mixture containing 10% of chlorine being heated to 40° C [20]; also by the interaction of chlorine being heated to $40°$ C [20] ; also by the interaction of acetylene and hydrogen chloride at ordinary temperature in the presence of an oxidising agent in a gaseous form, such as nitrogen peroxide, formic acid, formaldehyde, iodine, or oxygen, [21] and by passing a mixture of tetrachlorethane and steam over a heated iron or zinc catalyst. It should not be heated with alkalis, as this treatment causes the liberation of the spontaneously explosive substance, chloroacetylene. [22]

The characteristics of the pure isomers are as follows, [23] but there is some doubt which isomer is the cis and which the trans. [36]

	cis. trans.	cis.	trans.
sp. gr.		$1.265\ 1.2583_{20}\ 1.2502_{25}·289$	$1·2841_{20}\ 1.2771_{25}$
b. p. ($°$ C)		48.4	60.1
s. p. ($°$ C)		-50	-80
n_D		1.4490_{15}	1.4519_{15}
sp. ht. $15°-45°$ C)		0.295	0.295
lat. ht.		73.65	73.01
cub. exp.		0.00136	0.00127
visc.		$0.423_{20}\ 0.400_{25}$	$0.474_{20}\ 0.451_{25}$
s. ten.		$26.4_{20}\ 25.8_{25}$	$28.3_{20}\ 27.65_{25}$
flash p.		$39°$ F $(4° C)$	$43°$ F $(6° C)$
v. p.		280	183
dielec. const. $16°$ C		3.67	7.55

Four grades of industrial material are available: [23]

	sp. gr.	b. r. ($°$ C)	
1.	1.27	$48-59$	Mixture of isomers
2.	1.28	$54-62$	Mixture of isomers
3.	1.27	$57-53$	Mainly trans
4.	1.29	$58-62$	Mainly cis

Dichlorethylene is a solvent for some forms of cellulose acetate, also for ester gum, colophony, elemi, dammar, mastic, bitumen, rubber, rubber chloride, paraffin wax, seekay wax, sulphur, stearic acid, petroleum jelly, but not substantially for kauri, copals, pointianac, sandarac nor for shellac and tragacanth.

Trichloroethylene $CHCl = CCl_2$

Trichloroethylene is a clear, (non-inflammable) almost colourless liquid with a chloroform like odour. Commercially it is known as 'Triklone', (for industrial uses) or 'Trilene' (as an anaesthetic in childbirth). It has a TLV of 100 ppm.

It is immiscible with water to which it is stable but some decomposition is observed on being exposed to light or excessive heat. In its stabilised form it is stable and non-corrosive to most metals and in recent years a so-called neutrally (as opposed to amine or alkaline) stabilised solvent has been developed which enables it to be used without corrosive action on all metals, including aluminium and magnesium and their alloys.

Trichloroethylene is manufactured by dehydrochlorination of tetrachloroethane by boiling with lime or by thermal cracking.

It is decomposed by strong alkalis e.g. caustic soda and potash and silver oxide, to give the spontaneously flammable chloroacetylenes. Milder alkalies such as calcium hydroxide and sodium carbonate have no effect. It is also decomposed by hot concentrated nitric acid but not affected by cold hydrochloric and sulphuric acids.

The British Standard BS 580:1963 covers two types. Type 1 covers a stabilised material suitable for metal degreasing, dry cleaning, extraction of oils and fats and similar purposes. Type 2 provides a specially stabilised material suitable for unusually severe duty metal cleaning particularly for aluminium, magnesium and their alloys.

Table from BS 580: 1963

Type	1	2
Appearance	Clear and free from visible foreign matter	
Colour (Lovibonds units 6" cell)	0.4Y 0.1R	0.4Y 0.1R
sp. gr. $20/4^\circ$ C	$1.46-1.465$	$1.452-1.462$
b. r. 95% ($^\circ$ C)	$86.5-87.5$	$86.5-88.0$
Residue	0.005% w/w max.	0.005% w/w max.
Alkalinity as Na_2CO_3	$0.005-0.02$% w/w	0.0025% w/w max.
Free chlorine	Nil	Nil
HCl developed at 115° C	0.006% w/w max.	0.02% w/w max.

The pure material has sp. gr. $1.464_{20\ 4}$, $1.4556_{25\ 4}$; n_D 1.4776; b. p. 86.9° C; m. p. -87° C; sp. ht. $0.241_{19\ 42}$; lat. ht. 57.24; v. p. 36_{10}, 60_{20}, 94_{30}; K.B. value 130; visc. 2.10_{70}, 1.02_{30}, 0.71_0, 0.58_{20}; cub. exp. 0.00117; dielect const. 3.39 at 10_3 cps; thermal cond. 0.08; s. tens. 29.5_{20}, 29.0_{30}; crit. temp. 271° C; crit. p. 49.5 atm; crit. vol. 1.950 ml/ g; crit. dens. 0.513 gm/ml; vapour dens. 4.54 (Air = 1); ht. formation liquid at 25° C Hf $= -9.4$ Kcal/mol; sol. in water at trichloroethylene at 25° C 0.027% w/w

It forms the following azeotropic mixtures:

Trichlorethylene (%)	(%)	b. p. ($^\circ$ C)
64	methyl alcohol 36	60.2
73	ethyl alcohol 27	70.9
81	isobutyl alcohol 19	85.4
82	n-Propyl alcohol 18	81.8
86	isopropyl alcohol 14	74.0
69	ethanol 26. water 5	67.2
81	propanol 12. water 7	71.5
93	water 7	73.0

It is a solvent for coumarone, rubber, rubber chloride, bitumen, mastic, elemi, colophony, grease, many vegetable oils and alkaloids canada balsam, paraffin wax, seekay wax, stearic acid, lanoline, petroleum jelly, fluid silicones, ester gum, dammar, sulphur, polyvinyl acetate and methyl methacrylate resin, but will not dissolve cellulose nitrate, shellac or manila. It will dissolve some forms of cellulose acetate, especially in the presence of cyclohexanol. If mixed with methyl or ethyl alcohol it will dissolve ethyl cellulose. It does not dissolve polyvinyl chloride or acetochloride. It is a useful low-boiler for reducing the inflammability of lacquers. It dissolves DDT.

In addition to the specific solvent action referred to above trichloroethylene is used in metal degreasing, metal drying, textile scouring and finishing, oil and fat extraction and as a low temperature heat transfer medium. Other uses include preparation of resin in trichloroethylene for corrosion protection and lubricating films, trichloroethylene based paints, adhesives and many minor applications.

Perchloroethylene $CCl_2 = CCl_2$

This is a clear, almost colourless mobile non-flammable liquid miscible in all proportions with many organic solvents but immiscible with water. It has a TLV of 100 ppm. It is prepared from pentachloroethane by heating with lime or by direct chlorination of tetrachloroethane. Commercially known as 'Perklone' (in dry cleaning). It is stable to most metals (including Al and its alloys) but is discoloured on exposure to sunlight. It is decomposed by hot strong alkalis with explosive violence and reaction can occur in the cold.

BS 1593:1963 is summarised below and currently available grades of the solvent for dry cleaning, fat extraction, etc. meet this specification.

BS Table

Appearance	Clear almost colourless, free from suspended matter
Colour (Lovibond scale 6" cell)	0.6Y 0.1R
sp. gr. 15.5/15.5° C	1.63–1.635
b. r. 95%	120–122° C
residue	0.005% w/w max.
alkalinity as Na_2CO_3	0.002% w/w max. Neutral or alkaline to bromophenol blue
free chlorine	Nil
HCl developed after 48 hrs under reflux in presence of moist oxygen and steel strip	0.020% w/w max.
K.B.V.	90

Perchlorethylene forms the following azeotropic mixtures:

Perchlorethylene (%)	(%)	b. p. (° C)
19	ethyl alcohol 81	78
46	n-propyl alcohol 54	94
60	isobutyl alcohol 40	103
81	isoamyl alcohol 19	116
48	α-Epichlorhydrin 52	110

It is a colourless liquid, stable to moisture, but decomposes in the light, it does not attack metals.

Perchlorethylene dissolves amylose triacetate, bitumen, canada balsam, lanoline, paraffin wax, seekay wax, rubber, rubber chloride, stearic acid, sulphur, petroleum jelly, elemi, ester gum, dammar, colophony, mastic, but not cellulose esters, shellac, kauri, manila, pontianac, sandarac or tragacanth.

Trichlorotrifluoroethane $CCl_2 F CClF_2$

Known commercially as 'Arklone' this is a colourless non-flammable liquid of low toxicity with little smell. It is intrinsically stable to light, air, moisture and heat up to 300° C. It is manufactured by reacting perchloroethylene with chlorine and hydrofluoric acid in the presence of a catalyst.

The TLV is 1000 ppm.

The properties of the pure liquid are as follows.

sp. gr. at 25° C	1.565
b. p.	47.6° C
s. p.	−35° C
n_D at 25° C	1.3566
sp. ht.	0.22
lat. ht.	35.07
K. B. value	31
dielec const. at 25° C	2.35
surface tension at 25° C dyne cm^{-1}	19
therm. cond.	1.11×10^{-4}
solubility in water at 25° C	0.01% w/w
solubility of water in solvent	0.007% w/w
viscosity at 25° C	0.447

It is immiscible with water but miscible with most organic solvents. It dissolves cotton seed oil, cocoa butter and most mineral oils, paraffin wax, ester gum, colophony but not cellulose nitrate, shellac, rubber, PVC and other plastics. Because it has a selective solvent action dissolving oils and grease but not rubbers, plastics and varnishes its main application is that of a cleaning solvent for the electronic and electrical industry using ultrasonic and vapour cleaning methods to remove resin fluxes and other trace contaminants. Being a pure solvent containing no stabilisers or other additives it is used where high standards of cleanliness are vital and manufacturing specifications are particularly stringent. A typical analysis of commercial product is:

Purity as trichlorotrifluoroethane	99.8%
Other chlorofluorohydrocarbons	2000 ppm.
Boiling range	0.3 deg C max.
Residue	1 ppm max.
Moisture	15 ppm max.
Acidity (ppm as HCl) − free	0.5 max.
− by hydrolysis	0.5 max.
− by photolysis	0.5 max.

It forms azeotropes with:

Trichlorotrifluoroethane (%)		(%)	b. p. (° C)
93.6	methyl alcohol	6.4	39
96.2	ethyl alcohol	3.8	44.5
97.1	isopropyl alcohol [40]	2.9	46.5
87.5	acetone	12.5	43.6
91.0	chloroform	9.0	47.4
50.5	methylene chloride	49.5	36.5
60	trans-dichloroethylene [41]	40	44.5
99	water	1	44.5

The azeotropes are also offered as proprietary cleaning solvents to the electronic and electrical industries and for general cleaning purposes where selective solvent action is required.

Monochlorbenzene

Monochlorbenzene, or chlorbenzol, is a colourless mobile liquid, having a mild almond odour and a mild narcotic effect. It is prepared by the chlorination of benzene in the presence of chlorine-carriers, such as iron [29] or aluminium, also by the action of sulphuryl chloride in the presence of similar catalysts [30]. sp. gr. 1.112, 1.1065_{20}, 1.1013_{25}; b. p. $131.3°$; n_D 1.525; v. p. 8.7; flash p. 75° F (24° C); sp. ht. 0.308; lat. ht. 77.6; m. p. $-45°C$; elec. cond. below 1×10^{-9}; dielec. const. 5.6; visc. 0.801, 0.756_{25}; s. ten. $33.25'_{32}\,\dot{}_{65\,25}$.

British Standard Specification BS 2533:1954 for monochlorbenzene requires: sp. gr. 1.110–1.113 $(1.106-1.109)_{25}$; b. r. $131.3°-132.0°$ C, 95%; residue 0.01% max. not acid to bromphenol blue; free from free chlorine.

It is not miscible with water and does not develop acidity. It is miscible with most organic solvents and dissolves ethyl cellulose, soft copals, dammar, bakelite A, rubber, colophony, plextol, vinyl and other resins. It is useful for producing lacquers, which dry with a matt or 'crackle' surface.

It forms the following azeotropic mixtures:

Chlorbenzene (%)	(%)	b. p. (° C)
64	isoamyl alcohol 36	124.3
37	n-butyl alcohol 63	107.2
20	n-propyl alcohol 80	96.5

Sym-Dichlorethyl Ether $Cl . CH_2 . CH_2 . O . CH_2 . CH_2 . Cl$

Symmetrical or $\beta\beta'$-dichlorethyl ether is a colourless liquid, having an odour like that of ethylene dichloride (sym-dichlorethane). It is prepared by treating ethylene glycol chlorhydrine (β-chlorethyl alcohol) with sulphuric acid. Also as a by-product from glycol manufacture.

Characteristics: b. p. 178° C; m. p. −51.7° C; sp. gr. $1.219_{20\ 4}$; n_D 1.4573; v. p. 0.7; sp. ht. 0.369; lat. ht. 64; flash p. 131° F(55° C); ign. temp. 369° C; visc. 2.41, 2.14_{25}; s. ten. 37.6, 37.0_{25}; cub. exp. 0.001; it dissolves 0.28% of water at 20° and water dissolves 1.0% at 20° C, 1.7% at 90° C; K.B. value 40. It forms an azeotrope with 65.6% of water: b. p. 97.7° C. Good technical material [39] has sp. gr. 1.225–1.230; b. r. 5–95% within 3° C. including 177.5° C; flash p. 131° F (55° C).

It is a solvent for mastic, coumarone, colophony, benzyl abietate, ester gum, zinc resinate, paraffin wax, gum camphor, castor, linseed, cotton-seed and other fatty oils, turpentine, albertols, rezyls, polyvinyl acetate, chloride and chloroacetate, bakelite R 325, and glyceryl phthalate resins. It does not dissolve rubber, manila, dammar, kauri, congo, sandarac, pontianac, shellac, beeswax, carnauba wax or japan wax. It dissolves ethyl cellulose, but not other cellulose ethers or esters except in the presence of 10/30% of alcohol. It is miscible with aromatic but not with paraffin hydrocarbons.

It is not to be confused with symmetrical $\alpha\alpha'$-dichlorethyl ether, which is made by the action of hydrochloric acid on acetaldehyde. b. p. 116° C; sp. gr. 1.213_{20}; − 1.457.

1,2-Dichloropropane $CH_2Cl . CHCl . CH_3$

A colourless liquid known also as propylene dichloride. It has solvent and toxic properties similar to those of dichlorethane. sp. gr. 1.166 (1.158 at 20° C); b. p. 96° .4; n_D 1.439; flash p. 70° F (21° C) lat. ht. 72; sp. ht. 0.334; s. ten. 31; v. p. 36; m. p. −80° C; cub. exp. 0.0008; visc. 8; exp. lim. 3.4 to 14.5; ign. temp. 557°; therm cond. 0.073; dielec. const. 8.92.

It dissolves 0.04% of water at 20° C and water dissolves 0.27% at 20° C.

Dichlorpropane is a solvent for polyvinyl acetate and chloroacetate, colophony, ester gum, dammar, raw rubber, linseed oil and hydrocarbons. It will dissolve cellulose nitrate in the presence of alcohol. It does not dissolve cellulose acetate or shellac.

The isomeric 1,3-dichloropropane has sp. gr. $1.186_{20\ 4}$; b. p. 120.8; n_D1.448; visc. 1.034_{20}, 0.963_{25}; s. ten. 33.8.

Amyl Chloride $C_5H_{11}Cl$

A technical product is available [31] consisting of a mixture of:

	b. p. ($^\circ$ C)	sp. gr.	n_D
n-Amyl chloride	105	0.9013 (0°)	1.412
Isoamyl chloride	99	0.8928 (0°)	1.409
3-Chloropentane	103–105	0.916 (0°)	1.408
2-Chloropentane	97	0.912 (0°)	1.408
1-Chloro-2-methylbutane	97.6–99	0.8812 (17.5°)	–
tert-Amyl chloride	85	0.889 (0°)	1.405

It has sp. gr. 0.885; flash p. 38° F (3° C); n_D 1.406; cub. exp. 0.0011; visc. $_{25}$ 0.54; v. p. 43; K-B value 71; b. r. 85°–109° C. . .95%.

It is a solvent for waxes, oils, tars, rubber and resins, but not shellac, is non-miscible with water.

Pure n-amyl chloride has sp. gr. 0.884_{20}, 0.8795_{25}; b. p. 107.9° C; m. p. -99° C; n_D 1.4125; s. ten. 25; visc. 0.580, 0.547_{25}.

Dichloropentane

Also known as amylene dichloride is a light straw-coloured solvent for oils, resins, bitumens, raw rubber and vulcanised rubber. The industrial product [31] consists of about 40% of 1,4-dichloropentane and 60% of a mixture of the other dichloro derivatives of normal and iso-pentane. It has sp. gr. 1.06–1.08_{20}; b. r. 130°–200° C. . .95%; flash p. 106° F (41° C); ign. p. 130° F (54.5° C); n_D 1.448; sp. ht. 0.37; lat. ht. 68; forms a pseudo azeotrope with 34% of water. It is narcotic and tends to liberate hydrochloric acid.

References

[1] Brit. Pat 265259. *Cf.* B.P. 235044; 450373.
[2] U.S. Pat. 1482306 (1924).
[3] B.P. 458061. Usines de Melle.
[4] B.P. 445011; Soc. Carbochemique.
[5] *Asscn. Brit. Chem. Mfrs. Safety Summary*, 1930, **3**, 48.
[6] D.R.P. 222919.
[7] D.R.P. 129237.
[8] Brit. Pat. 116094.
[9] Brit. Pat. 245991.
[10] F.P. 327322; 355423.
[11] U.S. Pat. 992551; 1204608; 1260621; 1260622.
[12] Brit. Pat. 14709 (1915); U.S. Pat. 1009428; *Ind. Eng. Chem.*, 1919, p. 639.
[13] U.S. Pat. 870518 (1907).
[14] *Gas F.*, 1920, p. 691.
[15] Brit. Pat. 147908 (1920).
[16] *Lancet*, 1915, p. 544; *J. Med. Soc. Lond.*, 1915, **33**, 129; *Chim. et Ind.*, 1924, p. 763.
[17] *Chem. and Ind.*, 1921, p. 278A.
[18] Brit. Pat. 132757 (1919).

[19] Brit. Pat. 1105 (1912).
[20] D.R.P. 264006 (1912).
[21] U.S. Pat. 1540748.
[22] *Chem. and Ind.*, 1916, p. 451.
[23] Imperial Chemical Industries, Ltd.
[24] D.R.P. 274748 (1912).
[25] D.R.P. 263457 (1912).
[26] Brit. Pat. 132755 (1919).
[27] U.S. Pat. 1397134 (1921).
[28] Brit. Pat. 302321.
[29] *Compt. rend.*, **170**, pp. 1301, 1451.
[30] Brit. Pat. 259329.
[31] Sharples Chemicals Inc., Airco Export Co., N.Y., U.S.A.
[32] B.P. 302390. Färbenzeitung, 1941, p. 464.
[33] D.R.P. 436999 (1921).
[34] Brit. Pat. 513235.
[35] Brit. Pat. 519220.
[36] Mumford and Phillip. *J.C.S.*, 1950, p. 78.
[37] Albright and Wilson, Ltd., Oldbury, Warwickshire.
[38] Shell Chemicals, Ltd., London, W.1.
[39] Torkelson, T. R., Oyen, F., McCollister, D. D. and Rowe, V. K. 'Toxicity of
      1,1,1-Trichloroethane as determined on laboratory animals and human
      subjects'. *Am. Ind. Hyg. Ass. J. 1958* **19**, 353.
[40] J B.P. 1026003 ICI Ltd.
[42] B.P. 989155 ICI Ltd.

CHAPTER EIGHT

Furanes

Furfural

Furfural, known also as 2-furaldehyde and as 2-furan carbonal and sometimes inaccurately as furfurol, has been known for over a hundred years, having been discovered by Doebereiner in 1830, in the products of the destructive distillation of sugar. Its preparation from bran, straw, wood and other carbohydrate material is also a matter of history, and it was not until recent times that its manufacture on an industrial scale was undertaken. The raw material from which it is prepared is oat hulls, which was formerly a waste product obtained in the manufacture of rolled oats [1]. Also from bagasse, corncobs and cottonseed hulls which contain high proportions of pentosans. The hulls contain from 32 to 36% of pentosans, together with about 35% of cellulose and 10 to 15% of lignin; they are digested in carbon brick-lined rotating autoclaves, with dilute sulphuric acid. The autoclaves are fed with steam at 60 lb/sq. in., the steam carries off the furfural and small quantities of other volatile substances [10].

The steam-furfural vapour is condensed and passed into a column still from which a distillate is collected containing 65% of water and 35% of furfural; the distillate on condensation separates into two layers, the lower consisting of furfural with about 5% of dissolved water and the upper layer of water containing about 8% of furfural; this upper layer is returned to the column still for redistillation while the lower furfural layer is dried under diminished pressure to yield technical furfural of about 99.5% purity.

The dehydrated crude furfural can be decolourised and purified by vacuum distillation, the product slowly assumes its brown colour again, but the colouration is inhibited by the addition of 0.1% of hydroquinone or catechol.

Pure furfural has the following characteristics: sp. gr. 1.165, 1.160_{20}; n_D 1.5261; b. p. 161.7° C; s. p. −36.5° C; flash p. 133° F (56° C); ign. temp. 315°–357° C; s. ten. 49; vis. 1.49_{25}, 1.35_{38}, 1.09_{54}, 0.068_{99}; diel. const. 38 at 25° C; phenol coeff. 0.26. Technical furfural [1] has sp. gr. 1.161 at 20° C; b. r. 156°–167° C–99%. Residue 0.7%; ash 0.001%; water 0.02% max. Furfural mixes in all proportions with most organic solvents excepting petroleums, in which it is but slightly soluble. It is miscible with castor, chinawood, linseed and turky-red oils but only partially with other vegetable oils. It is not miscible with glycerol.

The mutual solubilities of furfural and water are given in the following table [7]:

Temperature °C	Furfural % weight in	
	Water layer	Furfural layer
10	7.9	96.1
20	8.3	95.2
40	9.5	93.3
60	11.7	91.4
80	14.8	88.7
97.9	18.4	84.1

The critical solution temperature [8] is 120.9° C when about 49.3% of water is present.

Furfural dissolves cellulose nitrate [3], acetate [4], and nearly all known cellulose esters and ethers, but is rather slow in its action; its solvent properties are increased by its admixture with alcohols and with benzene. It dissolves ester gum, coumarone, bakelites and glyceryl phthalate resins, but is not a good solvent for resins in general or rubber and is not compatible with gasoline.

Solutions of cellulose nitrate in furfural have viscosities approaching those in ethyl lactate and have but moderate toleration for dilution with toluene or butanol. The odour of furfural is strong and somewhat tiresome, although not definitely unpleasant; the vapour is slightly toxic in high concentrations, which are difficult to reach at ordinary temperatures. The toxicity of furfural is judged to about the same order as that of butyl alcohol. Furfural may cause allergic skin irritation.

The following azeotropic mixtures are known:

Furfural (%)	(%)	b. p. (° C)
38	pinene 62	143
55	cyclohexanol 45	156
40	camphene 60	147
36	water 64	98

Furfuryl alcohol. CH_2OH

This alcohol, known also as furyl carbinol, is a colourless, almost odourless, syrup obtained by the partial hydrogenation of furfural, under high pressure in the presence of base-metal catalysts; it tends to darken in colour on storage.

Pure furfuryl alcohol has sp. gr. 1.1296_{20}; n_D 1.485; b. p. 171° C; s. p. −20° C; s. ten. 38; flash p. 174° F (79° C).

Technical material has sp. gr. 1.137_{16}; n_D 1.485; b. r. 167°−177° C 95%; water 0.5% max. Furfuryl alcohol does not form an azeotrope with water.

It yields viscous solutions of cellulose nitrate [6], gels cellulose acetate and dissolves zein, ester gum, cumarone, sandarac, mastic, kauri, shellac, pontianac and phenolic resins. It does not dissolve congo, guaiac or dammar. It is miscible with water and many of the usual organic solvents with the exception of some petroleum hydrocarbons and vegetable oils. It attacks polythene but not nylon.

Furfuryl alcohol has some toxic action if ingested and prolonged contact with the skin should be avoided. It has been used industrially for many years without serious toxic effects.

Furfuryl acetate

This is obtained by the direct acetylation of furfuryl alcohol; it has the following physical properties: sp. gr. 1.1175_{20}; n_D 1.4627; b. p. 176° C.

It yields solutions of cellulose nitrate, the viscosities of which approximate to those in ethyl lactate, and which are moderately tolerant to dilution with aromatic and paraffin hydrocarbons. It is a solvent for ester gum and coumarone.

Tetrahydro furfuryl alcohol. CH_2OH

This is another product arising from the partial, but more complete, hydrogenation of furfural, it having been found possible, but not economic, to hydrogenate as far as n-amyl alcohol.

Tetrahydrofurfuryl alcohol is a colourless saturated substance and does not discolour with age like the unsaturated furfurals. It has a mild odour. Pure material has sp. gr. 1.0495_{20}; cub. exp. 0.0074; n_D 1.45;

b. p. 178° C; m. p. −105°/−110° C; s. ten. 25° C 36; visc. 6.3. Technical material has sp. gr. 1.052$_{25}$; n_D 1.45 at 25° C; b. r. 170–180° C 97%; flash p. 160° F (71° C). It is miscible with boiled linseed oil, blown soya oil, castor oil, water and most organic solvents. It does not form an azeotrope with water. Its physiological activity is low but it appears to have some action on the kidneys if ingested and is therefore unsuitable for foodstuffs. There seems to be no dermatitic risk.

It dissolves cellulose nitrate, yielding solutions having viscosities about double those of corresponding concentration in ethyl lactate; these solutions have high tolerances for aromatic hydrocarbons but are not readily compatible with the paraffins. Tetrahydrofurfuryl alcohol is a solvent for rubber chloride, ehtyl cellulose, benzyl cellulose, cellulose acetate, rosin, ester gum, coumarone, sandarac, pontianac, guaiac, vinyl acetate resin, mastic, kauri, soft copals, shellac, zein and water.

Furfural Derivatives of Less Importance [5, 7]

	sp. gr. at 20°C	n_D	b.p. (°C)
Methyl furoate	1.1786	1.4869	181
Ethyl furoate	1.1774	1.4782	197
n-Propyl furoate	1.0745	1.4750	211
n-Butyl furoate	1.0555	1.4740	233
Isoamyl furoate	1.0335	1.4720	232
Amylene glycol 1.2	0.9892	1.4413	210
Amylene glycol 1.5	0.9938	1.4499	237
n-Butylfurfuryl ether	0.955	−	190

The furoic or pyromucic esters all dissolve ester gum, coumarone, and cellulose nitrate, but not congo, guaiac, dammar, sandarac, mastic, kauri, pontianac or shellac. The solutions of cellulose nitrate have high viscosities, whilst their tolerance for toluene is about the same as that of diacetone alcohol; for butyl alcohol and paraffins it is high, excepting methyl furoate, which is incompatible with paraffins. The esters are all insoluble in water and tend to darken with age.

Furoic acid or pyromucic acid is obtained by the limited oxidation of furfural.

Furan [9]

Furan is made by the catalytic decarboxylation of furfural, it has also been synthesised industrially from acetylene. It is a powerful solvent and is miscible with most of the usual organic solvents. It has sp. gr.

0.9444, $0.937_{20\,4}$; n_D 1.4216; b. p. 32.1° C. It forms an azeotrope with 0.1% of water b. p. 30° C. Water dissolves 2.3% and it dissolves 0.2% of water at 20° C.

Tetrahydrofuran [9]

Tetrohydrofuran is made by the hydrogenation of furan which it resembles in olvent properties with the exception that it is soluble in water. It is also made by the liquid phase interaction of formaldehyde and acetylene in the presence of a copper-bismuth catalyst, followed by hydrogenation. It has sp. gr. 0.888_{20}; b. p. 66° C; $n_D 1.407°$ C; m. p. −65° C; flash p. 70° F (21° C).

Tetrahydrofuran seems to be somewhat narcotic and causes injurious effects on the kidneys and skin.

It forms an azeotrope with 5.4% of water b. p. 63° C.

Tetrahydrofuran tends to form a peroxide which renders distillation or evaporation dangerous.

References

[1] The Quaker Oats Company, U.S.A. Killeffer. *Ind. Eng. Chem.*, 1926, p. 1217. *Cf. Chem. and Ind.*, 1933, p. 608, 1947. B11 p. 135. B.P. 585772.
[2] Mains. *Chem. Met. Eng.*, 1922, p. 779.
[3] F.P. 472423 (1914).
[4] D.R.P. 307075 (1919).
[5] *Cf.* Trickey. *Ind. Eng. Chem.*, 1927, p. 643.
[6] F.P. 512850 (1931).
[7] The Miner Laboratories, Chicago.
[8] *Ind. Eng. Chem.*, 1926, p. 24.
[9] Imperial Chemical Industries and Du Pont de Nemours and Co.
[10] *Ind. Eng. Chem.*, 1948, p. 200. U.S.P. 1735084−1929.

CHAPTER NINE

Plasticising solvents

Acetophenone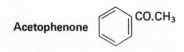

This substance is used as a softening plasticiser, it has powerful solvent properties, and a strong pleasant odour. It has a slight hypnotic action but it appears to be non-toxic.

Characteristics: The pure material crystallizes in plates or monoclinic prisms. m. p. 19.7° C and has sp. gr. 1.033 at 20° C; b. p. 202° C; n_D 1.535; lat. ht. 77; sp. ht. 0.434; visc. 1.8; flash p. 221° F (105° C); ht. comb. 989; ht. fusion 33; s. ten. 39.8; elec. cond. 6.43×10^{-9}; diel. const. 1.8; acid dissocn. const. 10^{-19}; dil. rat. toluene 2.7. Technical material has s. p. 17° C min; b. r. 90% within temperature range of 2.5° C including the temperature of 200° C.

It is a powerful solvent for cellulose nitrate and rapidly gels cellulose acetate. It also dissolves ethyl cellulose, shellac, coumarone, glyceryl phthalate resin, polystyrene, and vinyl resins. Polyvinyl chloride is either gelled or dissolved. It yields strong flexible and resplendent films with the cellulose esters. It is miscible with linseed and other oils and with hydrocarbons, but not with water.

Diacetin.
$$CH_2.OOC.CH_3$$
$$|$$
$$CH.OH$$
$$|$$
$$CH_2OOC.CH_3$$

Diacetin, or glyceryl diacetate, is a mixture of two isomeric esters of which the one shown above preponderates. It is occasionally employed when a soft film is required, especially in the case of cellulose acetate films, and also for the purpose of incorporating dyes in glyceryl phthalate resin.

It is a solvent for cellulose acetate and nitrate, shellac, glyceryl phthalate resin. It partly dissolves mastic, benzyl abietate, and castor oil. It does not dissolve ester gums, coumarone or copals, and is not miscible with linseed oil, paraffin hydrocarbons or toluene. It is miscible with benzene and with water and is hygroscopic.

British Standard Specification, BS 1594:1950 requires: sp. gr. 1.180–1.195 $(1.175–1.190)_{20}$; ash 0.02% max; acidity 0.4% max. as acetic; sulphates 0.05% max. as SO_4; chlorides 0.05% max. as Cl; esters 94–101%; clearly miscible with 2½ volumes of benzene. Diacetin also has n_D 1.44. b. p. about 259° C; flash p. 295° F (146° C); dil. rat. toluene 0.35. It usually contains monacetin and triacetin.

Triacetin.
$$
\begin{array}{l}
CH_2 . OOC.CH_3 \\
\mid \\
CH.OOC.CH_3 \\
\mid \\
CH_2 . OOC.CH_3
\end{array}
$$

Triacetin, or glyceryl triacetate, is one of the most widely used and valuable plasticisers. It possesses high plasticising power for cellulose nitrate, acetate, acetobutyrate and ethyl cellulose, for all of which it is a good solvent, and in which it can be incorporated in any proportion without causing cloudiness. It tends to yield soft films if used in proportions exceeding about 66% of the weight of the cellulose ester. It is particularly useful when it is desired to incorporate glyceryl phthalate resins, vinyl acetate and vinyl acetal resins, cyclohexanone formaldehyde resin or copal ester with mastic and hard copals, and is a moderate solvent for benzyl abietate, blown castor oil and linoxyn, but not rubber chloride. It is non-miscible with castor oil, linseed oil or petroleum hydrocarbons, but is completely miscible with aromatic hydrocarbons. It is non-hygroscopic.

Water dissolves 6.7% by weight. Triacetin dissolves 3.7% of water at 20° C.

British Standard Specification, BS 1997:1962 for triacetin requires: sp. gr.   1.160–1.170 $(1.156–1.166_{20}$, $1.152–1.162_{25})$; n_D 1.430–1.434; max. H_2O 0.10%; max. ash 0.02%; max. acidity 0.05% as acetic; esters 98.0–100.5% not acid to methyl orange; colour tests.

Its b. p. is about 258° C; m. p. −78° C; flash p. 270° F (133° C); n_D 1.431; visc. 22; dil. rat: cellulose nitrate 2.5 benzene, 0.9 toluene, 1.0 butanol; cellulose acetate 0.4 benzene, 1.3 ethanol, 0.7 butanol.

Glyceryl monolactate triacetate
$$
\begin{array}{l}
CH_2 . OOC.CH.(OOC.CH_3).CH_3 \\
\mid \\
CH.OOC.CH_3 \\
\mid \\
CH_2 : OOC.CH_3
\end{array}
$$

This substance known industrially as Plastil [3] is a pale yellow viscous liquid useful for plasticising cellulose acetate in proportions up to

150%. It is stable to light, has a fire-proofing effect and is practically non-volatile. sp. gr. 1.2; flash p. 320° F (160° C).

It is also a plasticiser for cellulose nitrate and is miscible with all the usual solvents but not with mineral oils or petroleum hydrocarbons. Water dissolves 3%.

$$CH_2.OOC.C_6H_5$$
Glyceryl tribenzoate. $CH.OOC.C_6H_5$
$$CH_2.OOC.C_6H_5$$

This solid plasticiser is an odourless, almost white, crystalline substance, and is soluble in all the usual solvents with the exception of petroleums. m. p. about 71° C; sp. gr. 1.25.

With cellulose nitrate it yields very hard, glossy but brittle films in proportions up to 150% by weight of the cellulose nitrate.

It has been used for cellulose acetate, with which it yields elastic films in proportions of up to 20%.

Mixed acetic and benzoic esters of glycerine have also been used as plasticisers [1].

Triphenyl phosphate. $(C_6H_5 . O)_3PO$

The plasticising properties of this ester are similar to those of tricresyl phosphate. It is a white crystalline solid substance, and is held by some to be unsuitable as a permanent plasticiser because of the possibility of it crystallising out of the film. This is only true if it be employed by itself in excessive proportions, but a second plasticiser, even in quite small quantities, will prevent it crystallising, a suitable substance being di-amyl phthalate. Triphenyl phosphate can be used with cellulose nitrate in proportions up to 100% of the weight of the latter, large doses tend to reduce the inflammability of films. It is not a satisfactory plasticiser for the acetate, and if used it should be mixed with diamyl phthalate, triacetin, benzyl alcohol or amyl tartrate. If the proportion of triphenyl phosphate used be in excess of 10% of the cellulose acetate it tends to yield cloudy films. It can be used in proportions up to 20% with cellulose acetobutyrate on which it exerts a small but definite protective action against ultra-violet light being better in this respect than the phthalates and sebacates. It is a plasticiser for diethyl cellulose, dibenzyl cellulose, cellulose stearate, laurate and oleate, polyvinyl acetate and shellac, the water-resistance of which it increases. It is a stable substance, and can be obtained commercially in a high state of purity.

British Standard Specification, BS 1998:1953 for triphenyl phosphate requires: s. p. $47°-49°$ C; ash 0.050% max; acidity 0.010% max. as phosphoric acid; tests for colour, stability and water solubility impurities.

b. p about $410°$ C; m. p. $48.5°$ C; completely soluble in amyl acetate, acetone, benzene. flash p. $450°$ F ($235°$ C); dil. rat., benzene 4.5, butanol 10.5; water dissolves 0.001% at $25°$ C.

Tricresyl phosphate. $(CH_3 . C_6H_4 . O)_3 . PO$

Known as tritolyl phosphate and TCP. This substance shares with triacetin and dibutyl phthalate the honour of being the most widely used plasticiser. There are no less than ten possible isomers of tricresyl phosphate, depending on the presence of ortho-, meta- and para-cresol in the cresol used. The isomer formerly used in commerce is the triorthocresyl phosphate, which is a crystalline substance of m. p. $18°$ C and sp. gr. $1.185-1.189$; n_D $1.560-1.562$.

This ortho variety is not now use; on account of its unexpectedly toxic nature [18], when ingested it can cause chronic paralysis of the legs. It has been replaced by the much less toxic product made from meta-para cresol which is known as ortho-free or m-p-tricresyl phosphate having sp. gr. $1.165-1.185$; n_D $1.554-1.560$; b. p. about $410°$ C; m. p. about $-35°$ C; pour point ASTM $-26°$ C; flash p. $446°$ F ($230°$ C); elec. cond. $1.5-2.0 \times 10^{-9}$; breakdown voltage (dry), $23-45,000$; visc. 265_{10}, 160_{15}, 120; s. ten. 45; dil. rat. toluene 3.3, benzene 4.5, butanol 10.5. The pure meta isomer is a liquid at ordinary temperatures and the pure para isomer has m. p. $78°$ C.

British Standard Specification, BS 1999:1964 for tricresyl phosphate requires: sp. gr. $1.163-1.17$($1.160-1.175_{20}$, $1.157-1.172_{25}$); n_D $1.554-1.559$; H_2O 0.10% max; acidity 0.020% max. as phosphoric acid; free phenols 0.05% max; tests for colour, colour stability and water-soluble impurities.

ASTM Specification, D 363-56(1965) requires: sp. gr. $1.150-1.180_{20}$; volatile matter 0.20% max; permanganate test 30 mins. min; H_2O 0.1%

max; unsaponifiable matter 0.5% max; acidity max. 0.1 mg KOH/g of sample: test for colour.

Tricresyl phosphate is an exceptionally stable ester; when quite pure it is practically non-volatile and odourless, and does not develop acidity, even under highly adverse conditions. It has a powerful solvent action on cellulose nitrate and acetonitrate, but little on cellulose acetate. It plasticises ethyl and benzyl cellulose, cellulose stearate, polystyrene, methyl methacrylate, polyvinyl chloride, acetate, chloracetate and acetal resins, rubber chloride and synthetic rubbers (up to 25%). It is a good solvent for ester gum, colophony, copal ester, benzyl abietate, coumarone, cyclohexanol-formaldehyde resin, glyceryl phthalate resin; it is a moderately good solvent for mastic and will dissolve hard copals on prolonged heating at high temperature. It serves as a good flux in which to run copals. It renders shellac waterproof and also improves its adhesive properties. It has some plasticising action on casein and gelatin. It is clearly miscible with castor, olive, nut, china wood, and linseed oils and with most of the hydrocarbons, but is non-miscible with water. It dissolves 0.2% of water at 25° C, and water dissolves 0.01% of tricresyl phosphate at 25° C.

Tricresyl phosphate diminishes the inflammability of cellulose nitrate films. It may be used in quantities up to 100% of the weight of cellulose nitrate, and up to 10% of cellulose acetate. It permits a high polish to be obtained and imparts flexibility to cellulose nitrate films without greatly decreasing the tensile strength being somewhat unique in this respect.

Opacity to uv light of over 290 nm can be imparted by means of tricresyl phosphate.

Tricresyl phosphate is a good plasticiser for polyvinyl chloride destined for use at a moderately low temperature, a mixture of 2 parts of polyvinyl chloride with 1 part of tricresyl phosphate has elec. cond. less than 1×10^{-14} and remains flexible at $-20°$ C (brittle temperature $-20/-25°$ C); the further addition of 1 part of dibutyl phthalate reduces the brittle temperature to $-47°$ C.

The stickiness sometimes engendered by ester gum may be counteracted with tricresyl phosphate.

The mixed esters, monophenyldicresyl phosphate and diphenylmonocresyl phosphate, have also been used as plasticisers but are not now in vogue but cresylxylyl phosphate (Tolylxylenyl phosphate) and trixylyl phosphate can be used to replace tricresyl phosphate to some extent.

Tolylxylenyl phosphate [13]

This is a clear, odourless, slightly brown oily liquid which is miscible with most of the organic solvents. It is a plasticiser for polyvinyl chloride and many of the synthetic rubbers. It has sp. gr. $1.135-1.146_{25}$; b. p. over $400°$ C; flash p. $500°$ F ($260°$ C); n_D $1.553-1.555$; visc. (centistokes) $90-125_{25}$, $4.6-5.0_{100}$.

Trixylenyl phosphate [8, 15] $(Me_2 . C_6H_3O)_3 . PO$

This is a water white oily liquid which is miscible with most of the organic solvents. It is a plasticiser for cellulose nitrate, acetobutyrate, ethyl cellulose, polyvinyl chloride. sp. gr. 1.15; b. p. $420°$ C; flash p. $529°$ F ($276°$ C); n_D 1.5516; visc. [15] (centistokes) 127_{20}, 54.4_{30}, 29.0_{40}, 17.7_{50}, 11.8_{60}, 8.6_{70}.

Triethyl phosphate. $(C_2H_5O)_3PO$

A colourless liquid having a mild odour resembling that of apples. It is miscible with most organic solvents and with water, which hydrolyses it slowly. It dissolves cellulose nitrate and acetate and tends to diminish inflammability. sp. gr. 1.076_{20}; n_D 1.406; b. p. $216°$ C; m. p. $-56°$ C; flash p. $240°$ F ($116°$ C); s. ten. 30, cub. exp. 0.000957.

Tributyl Phosphate. $(C_4H_9O)_3PO$

Tributyl phosphate is an odourless, colourless liquid, fast to light. It is a powerful solvent for cellulose nitrate and lowers the viscosity of its solutions; it also dissolves cellulose acetate, ethyl cellulsoe and shellac. It tends to cause cellulose nitrate films to develop acidity and brittleness on exposure. It is a plasticiser for neoprene, polyvinyl chloride and polyvinyl butyrol.

Characteristics: sp. gr. 0.989; n_D 1.424; b. p. $289°$ C (decomp); flash p. $295°$ F ($146°$ C); m. p. below $-80°$ C; lat. ht. 55; cub. exp. 0.00093; s. ten. 27.8; it dissolves 7.5% of water at $25°$ C and water dissolves 0.6% at $25°$ C; it diminishes the inflammability of cellulose nitrate; dil. rat. toluene 24.

Trichloroethyl phosphate. $(Cl . C_2H_4O)_3PO$

This is a non-inflammable plasticiser for cellulose acetate, nitrate. It is readily compatible with rubber chloride and butadiene acrylonitrile and to a useful extent with butadiene-styrene, polychloroprene, polyisobutylene, polyvinyl chloride and chloracetate. It is a colourless liquid of low viscosity. sp. gr. $1.42-1.46$; n_D $1.472-1.474$; b. p. $330°$ C; flash p. $437°$ F $(225°$ C); visc. 21, 1.3_{80}; acidity 0.01% max. HCl; Insoluble in water but soluble in all the usual organic solvents except petroleum hydrocarbons.

Trichlorophenyl phosphate. $(CL . C_6H_4 . O)_3PO$

A colourless liquid having sp. gr. 1.34; n_D 1.576. It is compatible with cellulose acetate in proportion up to 50%; less than 15% is sufficient to render it completely non-inflammable; it is also compatible with cellulose nitrate, polyvinyl chloride and firmvar. It imparts high water resistance to films.

It is miscible with aromatic hydrocarbons, alcohols, esters and ketones but not with petroleum hydrocarbons or low aromatic content.

Butyl oleate. $CH_3 . (CH_2)_7 . CH : CH . (CH_2)_7 . COOC_4H_9$

Butyl oleate is occasionally used for the purpose of imparting water-resisting and high-polishing properties to cellulose ester films; in this respect it is similar to amyl stearate, although less stable. It is necessary to employ in conjunction with it a plasticiser such as butyl phthalate for cellulose-nitrate or benzyl alcohol for cellulose-acetate films. When used alone the maximum quantity that can be incorporated is 5% of the weight of cellulose acetate and 2% of that of the nitrate. It is a yellow oily liquid which is not miscible with water. It does not dissolve cellulose acetate or nitrate, shellac or hard copals. It is a good solvent for ester gum, copal ester, coumarone, benzyl abietate, rubber chloride and is miscible with castor, linseed and other oils, also with hydrocarbons.

Characteristics: sp. gr. $0.87-0.88$; b. r. $350°-360°$ C (decomposes); m. p. about $-10°$ C; flash p. $356°$ F $(180°$ C); iodine value about 77; n_D 1.46.

Tetrahydrofurfuryl oleate $CH_3.(CH_2)_{16}.COO.CH_2$⟨O⟩

A light brown oil having a fatty odour. It is a plasticiser for polyvinyl chloride, is insoluble in water but miscible with most organic solvents.

Characteristics: sp. gr. 0.931; n_D 1.465; flash p. $374°$ F $(190°$ C); b. r. $260-285...90\%$.

Ethyl palmitate.

The technical product is a mixture of ethyl palmitate $C_{15}H_{31}$. $COO . C_2H_5$ and ethyl stearate $C_{17}H_{35} . COO . C_2H_5$. It has sp. gr. 0.857–0.859; b. p. approx. 340° C; esters 95% min as palmitate or 100–105% as stearate; acidity 0.1% max. as palmitic acid. It is a colourless liquid or a white soft wax having a compatibility of up to about 5% with cellulose esters and ethers, polyvinyl acetate, chloride and chloracetate and polyvinyl formals. It is used mainly for imparting water resistance and ease of polishing to films and silkiness to cellulose acetate threads, and as a lubricant for plastic extrusion processes.

Butyl stearate. $CH_3(CH_2)_{16} . COO . C_4H_9$

This is a stable ester, similar to amyl stearate, useful as a lubricating and water-repelling plasticiser; it permits a high polish and resistance to scratching to be given to cellulose ester films. The maximum permissible dose is 5% of the weight of cellulose acetate and 2% of that of cellulose nitrate. It should be used in conjunction with amyl phthalate, amyl tartrate or triacetin. m. p. 18°–20° C; sp. gr. at 20° C 0.855–0.860; n_D 1.4465; b. r. 355°–368° C; flash p. 293° F (145° C); cub. exp. 0.00083; acidity 0.2% max; water dissolves 0.2% at 25° C, and it dissolves 0.03% of water at 25° C. It also dissolves ester gum, copal ester, coumarone, benzyl abietate; it is miscible with oils and hydrocarbons, but not with water. It does not dissolve cellulose esters, shellac, hard copals, polyvinyl chloride, polyvinyl acetal, or glyceryl phthalate resins. It dissolves raw rubber on heating and increases the resistance of butadiene-styrene rubber to sunlight. It is occasionally used for plasticising ethyl cellulose and rubber chloride.

Amyl stearate. $CH_3(CH_2)_{16} . COO . C_5H_{11}$

This is a plasticiser best employed in conjunction with another plasticiser such as triacetin or amyl phthalate. Its value lies in its water-repelling action, and it allows a high polish to be given to films, 1% of the weight of the total solids being sufficient to impart these properties to films, the maximum proportion being 2% for cellulose nitrate and 5% for cellulose acetate. It is also valuable in lacquers incorporating copal esters, for which it is an effective plasticiser. It is useful as a plastic lubricant during moulding processes.

Characteristics: sp. gr. 0.870–0.880; b. p. about 360° C; m. p. 30° C; n_D 1.44–1.45; flash p. over 150° F (66° C).

It is a solvent for copal ester, coumarone, benzyl abietate; it partly dissolves ester gum, shellac and mastic, but not cellulose acetate or nitrate or vinyl acetate. It is miscible with castor, linseed and other varnish oils and with hydro-carbons, but not with water.

Cyclohexanyl stearate. $CH_3(CH_2)_{16}COOC_6H_{11}$

This material is an odourless white wax having m. p. $25°$ C; sp. gr. 0.885_{30}; esters 98%; acidity 0.08% max. It is insoluble in water, is stable to light and miscible with organic solvents and oils with the exception of methylated spirit. It is a plasticiser for polystyrene and its copolymer resins.

Methyl ethylene glycol stearate
$CH_3O . CH_2 . CH_2 . OOC . (CH_2)_{16} . CH_3$

Known industrially as methyl cellosolve stearate, this substance is a colourless oily liquid with a mild fatty odour.

Characteristics: sp. gr. 0.89_{25}; b. r. $320°-325°$ C; s. p. $21°$.

It is miscible with vegetable oils, hydrocarbons and all the usual solvents but not with water. It can be used for plasticising cellulose nitrate, ethyl cellulose and some resins.

Amyl benzoate. $C_6H_5 . COO . C_5H_{11}$

This ester is occasionally employed as a plasticiser; it is a colourless liquid having a mild odour. Sp. gr. $0.994-1.01$; n_D 1.4945; b. p. $262°$ C; cub. exp. 0.000858.

Amyl benzoate is a good solvent for ester gum, colophony, copal ester, coumarone, benzyl abietate, and is miscible with castor, linseed and other oils, also with hydrocarbons, but not with water. It is a poor solvent for cellulose nitrate and acetate, rubber chloride, shellac, and hard copals, and does not dissolve glyceryl phthalate resins.

It is possible to simulate permanently the effect of a white pigment in a cellulose acetate film by adding proportions in excess of 5% in the absence of another plasticiser.

Butyl benzoate. $C_6H_5 . COO . C_4H_9$

This is similar to amyl benzoate in properties; it has sp. gr. $1.009-1.010$; n_D 1.498; b. p. $249°$; m. p. $-22°$ C.

Benzyl benzoate. $C_6H_5 . COO . CH_2 . C_6H_5$

This is a very stable ester; it can be used in proportions up to 100% and 66% by weight of cellulose nitrate and acetate respectively. It yields strong hard films. It dissolves ester gum, colophony, copal ester, congo, kauri, Sierra Leone copals, mastic, dammar, coumarone, benzyl abietate, polystyrene, polyvinyl chloride, rubber chloride, butadiene acetonitrile, butadiene-styrene and methyl methacrylate resins and, on heating, hard copals. It has moderate solvent properties for shellac and glyceryl phthalate resins. It is miscible with castor and oils generally and with hydrocarbons, but not with water or glycerine. It is particularly of use when it is desired to incorporate a hard copal in a cellulose lacquer.

Characteristics: sp. gr. 1.126; b. r. 323°–324° C; n_D 1.568–1.569; m. p. 18°–21° C; flash p. 298° F (148° C); visc. 10; cub. exp. 0.00075; soluble in 2 volumes of industrial alcohol at 20° C; water dissolves 0.003%.

Butylene glycol dibenzoate
$CH_3 . CH(OOCCH_6H_5) . CH_2 . CH_2 . OOC . C_6H_5$

This plasticiser is an odourless, colourless light and water-stable liquid which imparts remarkable adhesive properties and high gloss to cellulose nitrate. sp. gr. 1.14; n_D 1.54–1.56; esters 98–100; acidity 0.2%.

In proportions up to 75% on cellulose nitrate and 30% on cellulose acetate; it yields clear, hard, flexible films and is a plasticiser for polyvinyl acetals.

Diethylene glycol dibenzoate
$C_6H_4 . COO . CH_2 . CH_2 . O . CH_2 . CH_2 . OOC . C_6H_4$

This substance is a solid, m. p. 15.9° C; sp. gr. 1.1765_{20}; n_D 1.5449; visc. 111.6; cub. exp. 0.00071; dissolves 1% of water. It is a plasticiser for cellulose nitrate, acetate, acetopropionate, acetobutyrate, polyvinyl alcohol, chloride, acetochloride, acetate, and butyral and ethyl cellulose.

Dimethyl phthalate. $\begin{array}{l}COO.CH_3\\COO.CH_3\end{array}$

An ester which is frequently used, it is similar to diethyl phthalate in properties, being a colourless, light fast odourless liquid. It is apt to leave films slowly in hot situations. Sp. gr. 1.193–1.197; b. p. 282° C; s. p. 0° C; n_D 1.515–1.517; ign. p. 132° C; flash p. 266° F (130° C);

Pour point ASTM $-26°$ C; visc. 17, 30_{10}; ht. comb. 5769; cub. exp. 0.00076; dil. rat. toluene 1.9; water dissolves 0.4%; dissolves 1.5% of water at $20°$ C. Dimethyl phthalate yields excellent films having good adhesion with both cellulose acetate and nitrate in proportions up to 75% of the weight of the cellulose ester. It is particularly stable to light. It is not miscible with petroleum naphtha and only partially miscible with turpentine and soya bean oil. It plasticises rubber and polyvinyl acetate and acetochloride but not polyvinyl chloride.

British Standard Specification, BS 1996:1962 for benzyl benzoate requires: sp. gr. 1.194–1.198 (1.191–1.195$_{20}$, 1.188–1.192$_{25}$); n_D 1.515–1.517; H_2O 0.10% max; ash 0.01% max; acidity 0.010% as phthalic acid; esters 99.0–100.5%; tests for colour and colour stability.

Diethyl phthalate.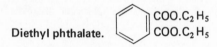

A stable ester known also [3] as diethyl 1,2-benzene dicarboxylate, enjoying a wide popularity for lacquer work. It is a good plasticiser for cellulose nitrate, cellulose acetate, acetobutyrate, acetonitrate, ethyl cellulose and dibenzyl cellulose, but is inferior to butyl or amyl phthalate for the nitrate. It can be used to plasticise cellulose acetate in proportions up to 100% by weight of the cellulose acetate and yields strong elastic resplendent films stable to light. The maximum proportionor cellulose nitrate is 80%, and for glyceryl phthalate resin 25%.

It is a good solvent for ester gum, colophony, coumarone, benzyl abietate, linoxyn, rubber chloride, glyceryl phthalate resin, polyvinyl acetate and cyclohexanone-formaldehyde resin; a moderate one for copal ester, shellc and mastic. Polyvinyl chloride absorbs about 5%. Hard copals can be dissolved in it on prolonged heating. It is miscible with castor, linseed and other oils and with benzene hydrocarbons; partly miscible with petroleums and non-miscible with water.

British Standard Specification, BS 574:1964 for dimethyl phthalate requires: sp. gr. 1.121–1.125 (1.118–1.122$_{20}$, 1.115–1.119$_{25}$); n_D 1.500–1.505; ash 0.01% max; H_2O 0.10% max; acidity 0.010% max. as phthalic acid; esters 99.0–100.5%; tests for colour and colour stability.

The material has b. p. about $295°$ C; visc. 20_{10}, 12.7; pour point ASTM below $-40°$ C; s. ten. 37; cub. exp. 0.00076; it dissolves 1.5% of water at $25°$ C and water dissolves 0.1% at $25°$ C; dil. rat. benzene 5.0, butanol 3.0, toluene 3.8; flash p. 284° F (140° C); lat. ht. 20.3 kG cal/mol. 20° C.

Diethyl phthalate is a good wetting medium in which to grind solid pigments.

Dibutyl phthalate. $COO.C_4 H_9$
$COO.C_4 H_9$

Dibutyl phthalate, known also [3] as dibutyl 1,2-benzene dicarboxylate, is one of the most widely used and satisfactory plasticisers for cellulose nitrate, ethyl cellulose and methyl methacrylate resin; it is also used for plasticising polystyrene, sulphonamide resins, rubber chloride, neoprene. GRN, thiokols, perbunan, nitrile and hycar synthetic rubbers. It is particularly good for polyvinyl chloride destined for electrical insulation at low temperatures, a mixture containing 33% of dibutyl phthalate has elec. cond. 1×10^{-12} and remains flexible at $-55°$ C (brittle temperature $-55°$ to $60°$ C) [6].

It is a solvent for ethyl and benzyl cellulose, polyvinyl acetate and chloroacetate, ester gum, copal ester, colophony, coumarone, benzyl abietate, mastic. It partly dissolves shellac, glyceryl phthalate resins and rubber. It dissolves hard copals and cellulose acetate on prolonged heating. It is miscible with castor, linseed and other oils and with hydrocarbons, but not with water. It can be used in quantities up to 100% of the weight of cellulose nitrate, but with cellulose acetate the maximum quantity permissible is less than 10%. In limited proportions it is a good plasticiser for glyceryl phthalate resins, polystyrene, polyvinyl butyral and rubber chloride.

British Standard Specification, BS 573:1964 for dibutyl phthalate requires: sp. gr. 1.049–1.053 (1.046–1.050 at 20° C); n_D 1.492–1.495; esters 99.0–100.5%; acidity 0.01% max; ash 0.01% max.

Dibutyl phthalate has b. p. about 335° C; s. p. −35° C; sp. ht. 0.43; cub. exp. 0.00074; lat. ht. 68; flash p. 329° F (164° C); it dissolves 0.3% of water at 25° C and water dissolves 0.04% at 25° C; dil. rat. benzene 4.0, toluene 3.8, butanol 20; visc. 70_0, 50_5, 38_{10}, 30_{15}, 25, 20_{25}; pour point ASTM −40° C.

ASTM Specification, D 608-58(1965) for dibutyl phthalate requires: sp. gr. 1.046–1.050 at 20° C; esters 99% min; acidity 0.01% max. as phthalic acid; clearly miscible with 19 volumes of 99% heptane at 20° C.

Diisobutyl phthalate $C_6H_4(COO . C_4H_9)_2$

A liquid resembling dibutyl phthalate but having a slightly higher solvent power. sp. gr. 1.049. b. r. 305°–315° C. flash p. 320° F (160° C).

Diamyl phthalate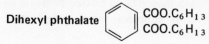

Diamyl phthalate is a substance with a very high boiling-point and of great stability, and it yields stable, lastic, weatherproof films of high tensile strength with cellulose nitrate. Diamyl phthalate can be used in proportions up to 100% with cellulose nitrate lacquer 10% with cellulose acetate, 50% with cellulose aceto butyrate. *Characteristics*: sp. gr. 1.025–1.027; b. r. 336°–342° C; m. p. below –55° C; n_D 1.488; esters 99–100%; acidity 0.025% max; dil. rat. benzene 3.0, toluene 2.3, butanol 20; flash p. 356° F (180° C); cub. exp. 0.00075; lat. ht. 25.2 k cal/mol. 20° C; it is practically odourless; water dissolves 0.04% at 20° C; dissolves 0.7% of water at 25° C.

It is a good solvent for ester gum, copal ester, polyvinyl chloride, acetate, chloroacetate and butyral, rubber chloride, methyl methacrylate, polystyrene, coumarone, benzyl abietate and mastic. Hard copals can be dissolved on prolonged heating. It has a limited solvent action on glyceryl phthalate resin and shellac. It is miscible with castor, linseed and other varnish oils and with hydrocarbons.

Dihexyl phthalate   COO.C_6H_{13}
                         COO.C_6H_{13}

There are two industrial varieties of dihexyl phthalate, viz:

1. Di-hexyl phthalate made from hexan-1-ol (normal hexyl alcohol).
2. Diethylbutyl phthalate made from 2-ethylbutan-1-ol (diethyl ethanol).

They resemble one another closely and are remarkably stable esters

Characteristics:

	n-hexyl	Ethylbutyl
sp. g.	1·01–1·02	1·019
n_D	1·489–1·493	1·492
visc.	29	78_{15} 40_{25}.
m.p.	–58°	–
flash p.	375° F.	–
cub. Exp.	0·00077	–
water dissolves	0·24%	–·

withstanding high temperatures without decomposition and low temperature without crystallising.

They are good plasticisers in proportions up to 100% for cellulose nitrate, porpionate, butyrate, ethyl cellulose, methyl methacrylate, polyvinyl chloride, chloroacetate, butyral, polystyrene and for cellulose acetate up to 15% and polyvinyl acetate up to 10%. They dissolve dammar, cumarone, and to some extent shellac and ester gum.

Dioctyl phthalates 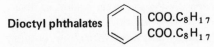 $COO.C_8H_{17}$
$COO.C_8H_{17}$

There are three industrial varieties of di-octyl phthalate:

1. Di-octyl phthalate made from octan-1-ol (normal primary octyl alcohol) and called DNOP.
2. Di-octyl phthalate made from octan-2-ol (capryl alcohol) and sometimes called dicapryl phthalate or DCP.
3. Diethylhexyl phthalate made from 2-ethylhexan-1-ol (ethylhexyl alcohol) and called DOP.

These three plasticisers resemble one another closely and in general can replace one another; they are pale yellow, almost odourless viscous liquids, miscible with most of the usual solvents, stable to light, heat and water, have no action on copper and of negligible toxicity.

Characteristics:

	n-Octyl	s-Octyl	ethylhexyl
sp. gr.	0.978–0.990	0.971–0.978	0.985–0.090
b. p. ($^\circ$ C)	c 340	c 325	c 386
m. p. ($^\circ$ C)	−40	below −60	−46
flash p. ($^\circ$ C)	219	201	210
visc.	40	69	81
n_D	1.485–1.488	1.480–1.488	1.486
lat. ht.	−	−	55
cub. exp.	−	−	0.00076
elec. cond. 25° C	−	−	1×10^{-6}
water diss. 20°	<0.02%	<0.03%	<0.01%
diss. water	0.6%	−	0.2%

The solvent and plasticising properties of these three varieties are generally similar and are as follows. Cellulose nitrate is miscible in all proportions, the plasticising proportion in 33–40%. Cellulose aceto-butyrate has a maximum compatibility with about 15% yielding good moulding powders but weak films with high water resistance. Polyvinyl chloride is miscible in all proportions yielding films resistant to both low and high temperatures. A mixture containing 33% of plasticiser yields films which are still flexible at $-35°$ C (brittle temperature $-35°$ to $-40°$ C) and which have elec. cond. 8×10^{-12}. Polyvinyl acetochloride is miscible in all proportions, 20—40% of plasticiser gives fungicidal films which have high tensile strength and tear resistance. These plasticisers are highly compatible with ethyl cellulose, polythene, polystyrene, butadiene-styrene, polychloroprene. The plasticising proportion for methyl methacrylate is 25% and for butadiene-acrylonitrile 25—45% of plasticiser. They dissolve ester gum, dammar, coumarone and partially shellac but they are not compatible to any useful extent with cellulose acetate, polyvinyl formal or butyral, polyvinyl acetate, cellulose acetopropionate, vinylidene-vinyl chloride, isoprene-isobutylene.

British Standard Specification, BS 1995:1962, for di-(2-ethylhexyl) phthalate requires: sp. gr. 0.985—0.991 $(0.983—0.989_{20}$, $0.981—0.987_{25})$; n_D 1.484—1.490; H_2O 0.10% max; ash 0.01% max; acidity 0.025% as phthalic acid; esters 99.0—103.0% as di-(2-ethyl-hexyl) phthalate; tests for colour and colour stability.

This substance is also known as di-2-ethylhexyl-1,2-benzene di-carboxylate.

Dilauryl phthalate

COO.$(CH_2)_{11}$.CH_3
COO.$(CH_2)_{11}$.CH_3

Dilauryl phthalate is the phthalate of the higher saturated aliphatic alcohol, n-dodecanol sometimes termed C^{12} alcohol. It is a neutral pale yellow almost odourless liquid of very high boiling-point, having sp. gr. 0.9459 $(0.9433_{204}$; n_D 1.482. It imparts resistance against the prolonged action of light and moderate heat on cellulose nitrate films; in proportions up to 66% it is a good plasticiser for cellulose nitrate, ethyl cellulose, polystyrene, polyvinyl chloride, chloroacetate and acetals.

Methylcyclohexyl phthalate [3]

COO.C_6H_{10}.CH_3
COO.C_6H_{10}.CH_3

A pale amber coloured viscous liquid of mild odour and very low volatility. It is a mixture of the phthalates of the three isomeric methyl cyclohexanols. It has sp. gr. 1.075; visc. 108_{25}; it is stable to light and heat and miscible with all the usual organic solvents but not with water.

It is a plasticiser for cellulose nitrate and permits the incorporation of cumarone resins. It plasticises shellac, polyvinyl chloride, polystyrene, rubber chloride and urea-formaldehyde resin. Cyclohexyl phthalate is a solid plasticiser, m. p. 64°–65° C similar in properties to methyl cyclohexanyl phthalate.

Dimethylglycol phthalate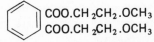

Known also as ethylene glycol mono methyl ether phthalate, di-2-methoxy ethyl phthalate and as methyl cellosolve phthalate, is a plasticiser for both cellulose acetate and nitrate in proportions up to 50% of the weight of the cellulose ester, for shellac up to 25%, rubber chloride up to 20% and polyvinyl chloroacetate up to 100%. It is a liquid, having sp. gr. 1.17; n_D 1.504; b.p. 310°C approx; flashp. 345° F (174° C); dissolves 3.4% of water at 20° C; water dissolves 0.02% at 25° C; gasoline dissolves 2% at 20° C; visc. $26_{25}, 4_{80}$.

Diethylglycol phthalate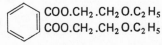

This substance is the phthalic ester of ethylene glycol monoethyl ether, known also as diethoxyethyl phthalate and as cellosolve phthalate; it is an excellent softener for cellulose nitrate, and it gelatinises cellulose acetate in proportions up to 50%. It is a colourless crystalline solid. m. p. 34° C; sp. gr. 1.123; n_D 1.492; b. p. 345° C; flash p. 289° F (143° C). Soluble in alcohol, acetone, benzene, insoluble in petroleum hydrocarbons. Water dissolves 0.2%. It tends to cause the darkening of cellulose nitrate films.

Dibutylglycol phthalate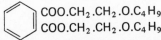

Ethylene glycol monobutyl ether phthalate is a liquid which boils at 370° C with slight decomposition, it can be used in proportions up to 33% of cellulose acetate, also with rubber chloride up to 20% and polyvinyl chloroacetate up to 100%. It also plasticises ethyl cellulose and methyl methacrylate. It is soluble in practlcally all organic solvents. sp. gr. 1.065 at 20° C; visc. 32; n_D 1.488; m. p. below −50° C; flash p. 406° F (208° C); s. ten. 33; sp. ht. 0.54.

Diphenyl phthalate. COOPh
                         COOPh

A white odourless solid, m. p. 69° C; sp. gr. 1.28 at 25° C; b. p. 405° C; discolours in light. Its maximum compatibility with cellulose nitrate is 80%, with acetate 50%, with gum dammar 30%; its is also compatible with ethyl and benzyl cellulose, polystyrene and vinyl resins.

This plasticiser tends to produce hard inflexible films having high gloss and weather resistance.

The following figures [5] show the solubility of diphenyl phthalate in grams per 100 cc at 25° C.

Methanol 12, ethanol 9, butanol 4, acetone 215, methyl ethyl ketone 140, ethyl acetate 95, amyl acetate 25, ethylene dichloride 180, trichlorethylene 120, carbon tetrachloride 70, benzene 90, xylene 45, monochlorbenzene 110, turpentine 10, pine oil 22, tuing oil 13, raw linseed oil 12, soya bean oil 10, castor oil 10, mineral oils 0, water 0.

Benzyl butyl phthalate [8] COO.CH$_2$.Ph
                                COO.C$_4$H$_9$

An almost colourless oil of slight odour, miscible with all the usual solvents and diluents but insoluble in water. It has sp. gr. 1.11 approx; n_D 1.533 approx.; it is compatible with cellulose nitrate, acetopropion-ate, acetobutyrate, polyvinyl chloride and butyral, rubber chloride, polystyrene and some acrylic resins. It is not compatible with cellulose acetate. It is stable to light, cold and heat and very good for extrusions and hot spraying processes.

Benzyl methyl phthalate is also available [8]. It has sp. gr. 1.18; n_D 1.555. Its properties are similar to those of benzyl butyl phthalate.

Diethylhexyl tetrahydrophthalate [16] COO.CH$_2$.CH(C$_2$H$_5$).C$_4$H$_9$
                                           COO.CH$_2$.CH(C$_2$H$_5$).C$_4$H$_9$

This somewhat viscous oil has sp. gr. 0.968$_{20}$; n_D 1.465; s. p. about −50° C; visc. 42; flash p. 350° F (177° C); cub. exp. 0.00077; max. power factor 25°−60 cy. 8%. It is completely compatible with cellulose nitrate, polyvinyl chloride, acetochloride and butyral; about 10% compatible with ethyl cellulose, polyvinyl acetate, methyl methacrylate but not with cellulose acetate or acetobutyrate. It imparts good low temperature flexibility without loss of good electrical properties.

$$COO.CH(C_2H_5).C_4H_9$$
$$|$$

Di-2-ethylhexyl adipate [8, 16, 19] $(CH_2)_4$
$$|$$
$$COO.CH(C_2H_5).C_4H_9$$

This liquid plasticiser is particularly useful for natural and some of the synthetic rubbers and polyvinyl chloride. It has sp. gr. 0.928–0.932 $(0.924–0.930)_{20}$; n_D 1.448; s. p. about $-60°$ C; flash p. $390°$ F $(199°$ C); visc. 13.7; cub. exp. 0.00081; dissolves 0.12% of water. It is completely miscible with polyvinyl chloride, chloroacetate and ethyl cellulose about 25% with cellulose acetopropionate, acetobutyrate, cellulose nitrate and polyvinyl butyral, 10% with methyl methacrylate but not with cellulose acetate or polyvinyl acetate.

Dinonyl adipate is also available [8]. It has sp. gr. 0.9185–0.9195; n_D 1.447–1.448.

$$COO.C_4H_9$$
$$|$$

Dibutyl sebacate. $(CH_2)_8$
$$|$$
$$COO.C_4H_9$$

An almost colourless liquid plasticiser for cellulose nitrate, acetopropionate and acetobutyrate, ethyl cellulose, rubber, butadiene acrylonitrile, polychlorobutadiene, nitrile rubber, buna N, thioplasts, rubber chloride, and polyvinyl chloride, acetochloride and butyral. It is of special value for electrical insulation with this last material when required for use at low temperature. A mixture containing 33% of dibutyl sebacate has elec. cond. 5×10^{-10} and is still flexible at $-70°$ C (brittle temperature $-70°$ C to $-75°$ C). It is not compatible with cellulose acetate or polyvinyl formal.

Characteristics: sp. gr. 0.940–0.945; n_D 1.44–1.45; m. p. $-8°$ C; b. p. about $345°$ C; flash p. $356°$ F $(180°$ C); ign. p. $410°$ F $(0°$ C); visc. 8.7_{25}, 2.9_{80}; lat. ht. 24.4 k.cal. mol $20°$ C; dielec. const. 3.6; power factor 6 (60 cyc.); water dissolves 0.01% and it dissolves 0.2% of water at $25°$ C.

British Standard Specification, BS 2535:1962, for di-n-butyl sebacate requires: sp. gr. 0.939–0.944 $(0.936–0.941_{20}$, $0.933–0.938_{25})$; n_D

1.441–1.445; H_2O 0.10% max; ash 0.01% max; acidity 0.025% max as sebacic acid; esters 98.5–101.0% as di-n-butyl sebacate; tests for colour and colour stability.

Dioctyl sebacate [8, 13]
$$COO.CH_2.CH(C_2H_5).C_4H_9$$
$$(CH_2)_8$$
$$COO.CH_2.CH(C_2H_5).C_4H_9$$

Known more correctly as di-(2-ethylhexyl) sebacate has sp. gr. about 0.92_{20}; n_D 1.45; m. p. $-50°$ C approx.; flash p. $415°$ F ($213°$ C); visc. 23_{25}, 5.3_{80}.

British Standard Specification, BS 2536:1962 requires: sp. gr. 0.917–0.923 ($0.915–0.921_{20}$, $0.913–0.919_{25}$); n_D 1.450–1.454; H_2O 0.10% max; ash 0.01% max; acidity 0.025% max. as sebacic acid; esters 99.0–101.0% as di-(2-ethylhexyl) sebacate; tests for colour and colour stability.

It is particularly good with ethyl cellulose, rubber chloride and polyvinyl chloride with the last a dose of 33 to 40% gives films which are still flexible at about $-60°$ C. It is a moderately good plasticiser for cellulose nitrate, cellulose acetobutyrate, polyvinyl chloroacetate and the butadiene-acrylonitriles but is poor with polyvinyl acetate, cellulose acetate, shellac and butadiene-styrene.

It is stable to light and heat and soluble in all the usual solvents and diluents.

Triethyl citrate.
$$CH_2.COO.C_2H_5$$
$$C(OH).COO.C_2H_5$$
$$CH_2.COO.C_2H_5$$

This is a plasticiser for both cellulose acetate, acetobutyrate and nitrate, with which it can be used in proportions up to 100% by weight of the cellulose ester. The pure ester has b. p. $294°$ C; sp. gr. 1.140–1.146; n_D 1.445; dil. rat. benzene 4.0, toluene 3.9, xylene 2.0, butanol 4.0; flash p. $155°$ F ($68°$ C); visc. 28, 3_{80}. It is a colourless, odourless mobile liquid having a very bitter taste [14]. It dissolves ester gum, dammar, shellac, glyceryl phthalate resin, some albertols, ethyl cellulose and polyvinyl acetate. It is a softener for starch ethers. Water dissolves 6.5%.

Tributyl citrate

$$CH_2.COO.C_4H_9$$
$$|$$
$$C(OH).COO.C_4H_9$$
$$|$$
$$CH_2.COO.C_4H_9$$

A liquid plasticiser having a weak, fruity odour. It plasticises both cellulose nitrate and acetate, the maximum permissible proportions being 75% and 20% respectively. sp. gr. 1.048–1.053; n_D 1.446; m. p. −20° C; flash p. 360° F (182° C); visc. 34, 4.3_{80}; dil. rat. benzene 6.8, toluene 4.9, butanol 19.0, petroleum 1.9. It may contain small proportions of the dibutyl ester of acetone dicarboxylic acid. When used in excessive proportions with cellulose acetate it tends to cause lack of adhesion, but within the range given above it improves the adhesion of cellulose nitrate films and resistance to oils and greases. It imparts high tensile strength at high elongation. It is readily compatible with ethyl cellulose, polyvinyl chloride, acetate, acetochloride and butyral, cumarone, shellac, dammar and ester gum. xylene 2.0, butanol 4.0; flash p. 155° F (68° C); visc. 28, 3_{80}. It is a colourless, odourless mobile liquid having a very bitter taste [14]. It dissolves ester gum, dammar, shellac, glyceryl phthalate resin, some albertols, ethyl cellulose and polyvinyl acetate. It is a softener for starch ethers. Water dissolves 6.5%.

Triamyl citrate

$$CH_2.COO.C_5H_{11}$$
$$|$$
$$C(OH).COO.C_5H_{11}$$
$$|$$
$$CH_2.COO.C_5H_{11}$$

This is similar in character to tributyl citrate; it has sp. gr. 1.013; n_D 1.447; esters 97–100%; acidity 0.1% max; dil. rat. benzene 5, xylene 6, butanol 11. It is a good solvent for cellulose nitrate. ester gum, glyptals, dammar, shellac, soluble bakelites, but not for cellulose acetate.

Dibutyl tartrate

$$CH(OH)COO.C_4H_9$$
$$|$$
$$CH(OH)COO.C_4H_9$$

Butyl tartrate, like amyl tartrate, is particularly suitable for use with cellulose acetate, ethyl cellulose and glyceryl phthalate resins, being employed in quantities up to 100% of the weight of these, and up to 150% with cellulose nitrate. It yields stable weatherproof films.

Characteristics: sp. gr. 1.090–1.095, (1.075–1.09 at 20° C); b. p. 298° C; n_D 1.45; flash p. about 320° F (160° C); m. p. (pure) 21.8° C; Visc. 97; dissolves 4% of water at 25° C and 3% at 180° C; water dissolves 1% at 25° C; it is somewhat easily hydrolysed by water; dil. rat. (cellulose nitrate): benzene 13.5, toluene 10.7, butanol 20; (cellulose acetate): benzene 1.2, ethyl alcohol 1.0, butanol 0.2.

It has good compatibility with cellulose nitrate and acetate, colophony, ester gum, copal ester, benzyl abietate, sulphur, glyceryl phthalate resin, polyvinyl acetate, acetochloride and acetals, methyl methacrylate resin, shellac, coumarone, and–in conjunction with benzene and methyl acetate–for ethyl cellulose. It has limited compatibility with polyethylene, polystyrene, polyvinyl chloride, and polymers of styrene with butadiene or acrylonitrile. It is miscible with oils and hydrocarbons. It can be used with bakelite resins, rubber chloride, zein and caseine.

Diamyl tartrate
$$CH(OH).COO.C_5H_{11}$$
$$|$$
$$CH(OH).COO.C_5H_{11}$$

This is an exceptionally good plasticising solvent for cellulose acetate and for glyceryl phthalate resin. It enables these two substances to be effectively combined so as to give resplendent, stable, weatherproof films; the maximum proportion recommended is 80% of the weight of the cellulose acetate. The maximum dose for cellulose nitrate is 100%.

Physical Characteristics: sp. gr. 1.04–1.06; n_D 1.45; b. p. about 400° C; dil. rat. toluene 9.8; flash p. over 200° F (93° C). At 25° C water dissolves 1% and it dissolves 4% of water.

It is a good solvent for cellulose nitrate and acetate, colophony, ester gum, bakelite resin, glyceryl phthalate resin, coumarone, benzyl abietate, shellac, rubber chloride, polyvinyl acetate, chloride, acetochloride, formal and butyral.

It has moderate solvent powers for copal ester, mastic, soft copals, and will not dissolve hard copals. It is miscible with castor and linseed oils and hydrocarbons. In conjunction with benzene and methyl acetate it dissolves ethyl cellulose.

Methylcyclohexyl oxalate $(COO.C_6H_{10}.CH_3)_2$

Methylcyclohexyl oxalate is a plasticiser known industrially as 'Barkite' [1]. It is a viscous, practically colourless and odourless liquid consisting of a mixture of three isomerides. sp. gr. 1.03; esters 98–100%, acidity

0.02% max; visc. 37 at 25° C; flash p. 293° F (147° C); insoluble in water; anhydrous; miscible in all proportions with the usual lacquer solvents and diluents, castor and drying oils.

It is a solvent for cellulose nitrate, ethers, coumarone, indene and other resins; it has good wetting power for pigments and is stable to light and heat. It causes rubber to swell.

The maximum proportions for plasticising cellulose acetate and nitrate are 66% and 150% respectively.

The homologous *cyclohexyl oxalate* is a white crystalline solid, m. p. 45° C, having a faint odour. It forms a solid solution with cellulose nitrate and is compatible with resins.

Dimethylcyclohexyl oxalate $(COO . C_6H_9 . (CH_3)_2)_2$

Known industrially as Barkite B' [1] is a substance similar to methylcyclohexyl oxalate and has sp. gr. 1.016; visc. 44 at 25° C; esters 98% min; acidity 0.02% max; flash p. 307° F (153° C).

Butylene glycol monolactate
$CH_3 . CH(OH) . CH_2 . CH_2 . OOC . CH(OH) . CH_3$

This substance is a non-volatile pale yellow viscous liquid which, although completely soluble in water, yields clear, hard, flexible water-resisting films with both cellulose nitrate and acetate in proportions up to 85% of the weight of the cellulose ester. It has sp. gr. 1.09–1.10; n_D 1.44–1.45; miscible with nearly all solvents.

Monocresylglyceryl ether
$CH_3 . C_6H_4 . O . CH_2 . CH(OH) . CH_2OH$

An almost odourless solid having the following characteristics [8] : sp. gr. 1.14–1.15; n_D 1.535–1.540; b. r. 315°–330° C; m. p. 30°–50° C; solubility in water at 15° C 1–1.5%; dissolves 10% of water at 15° C; Acetyl value 400–425; non-hydrolysable, neutral; miscible in all proportions with all the usual solvents excepting paraffin hydrocarbons. Plasticises benzyl cellulose, ethyl cellulose and urea-formaldehyde resins [9] but not cellulose acetate or nitrate; inert to vulcanised rubber.

Dicresylglyceryl ether
$CH_3 . C_6H_4 . O . CH_2 . CH(OH) . CH_2 . O . C_6H_4 . CH_3$

Similar to monocresyl ether and has the following characteristics [10] : sp. gr. 1.136; n_D 1.549; b. r. 328–340° C; insoluble in water; dissolves 10% of water at 15° C; acetyl value 220–260; plasticises benzyl and ethyl cellulose; soluble in all the usual organic solvents; inert to vulcanised rubber.

$$CH_2O.C_6H_4.CH_3$$
$$|$$
Monocresylglyceryl ether diacetate $CH.OOC.CH_3$
$$|$$
$$CH_2.OOC.CH_3$$

A liquid plasticiser for cellulose acetate, cellulose nitrate, ethyl cellulose and benzyl cellulose. This substance is much more resistant to hydrolysis than triacetin, and when used in proportions of from 25–50% of the weight of the cellulose ester or ether, it yields non-oxidising flexible and permanent films; the maximum permissible dose is 100%. It dissolves ester gum and is soluble in all the usual solvents and diluents.

Characteristics [8] : b. p. about 310° C; sp. gr. about 1.35; n_D about 1.49.

$$CH_2.O.C_6H_4.CH_3$$
$$|$$
Dicresylglyceryl ether monoacetate $CH.OOC.CH_3$
$$|$$
$$CH_2.O.C_6H_4.CH_3$$

This substance has properties similar to monocresylglyceryl ether diacetate, but is still more resistant to hydrolysis and has a higher boiling-point.

Characteristics [18] : b. p. about 360° C; sp. gr. about 1115; n_D about 1.53.

Camphor

The doyen of plasticisers, has been largely replaced by more suitable substances, but is still used for celluloid in proportions up to 25%. Camphor by itself has no apparent action on cellulose nitrate, but in the presence of quite small proportions of methyl, ethyl, butyl or amyl alcohols, its solvent action is rapid. It stabilises cellulose nitrate and reduces the explosivity but not the inflammability of this substance. It is useless with cellulose acetate.

As a plasticiser, it has three serious defects; its strong odour, its relatively high vapour pressure which leads to impermanence, while excessive proportions of it tend to cause wax-like films.

Characteristics: sp. gr. 0.990–0.995; m. p. 175–176.6° C, (pure) 178.7° C; b. p. 204° C sublimes; flash p. 158° F (70° C); lat. ht. 93; One part of camphor is soluble in 700 of water, or in 0.7 parts of 95% alcohol at 15° C; it readily dissolves in most organic solvents.

It is mainly produced by physical methods from Japanese camphor oil and by chemical synthetical methods from turpentine and other essential oils, the chief difference between the natural and synthetic products being the optical rotation, which for natural camphor is about +44° C. Synthetic camphor is as efficient as the natural for plasticising cellulose nitrate.

Castor oil

Castor oil is expressed from the seeds of *Ricinus communis*, and consists mainly of the glycerine esters of the following acids in approximately the proportions given: ricinoleic 80%, oleic 9%, stearic 3% and linoleic 3%. The characteristics of normal oils are:

British Standards Specification, BS 650:1967, for 'Firsts' quality Castor Oil requires: sp. gr. 0.958–0.969; n_D 1.477–1.481; Wijs iodine value 82–90; saponification value 177–187; acidity 2.0% max. as oleic acid; acetyl value 140 min; unsaponifiable matter 1.0% max; test for colour.

Castor oil is soluble in most organic solvents with the exception of methyl alcohol, triacetin, diacetin and some paraffin hydrocarbons.

Castor oil is only soluble in cellulose nitrate in very limited proportions; with most forms of cellulose nitrate, and in the absence of other substances, it gives clear films with proportions up to about 5%. Larger quantities give rise to films in which the oil is dispersed in microscopic particles, in which state it exerts a gelatinising lubricating effect. In proportions over about 25%, the microscopic particles of oil coalesce and exude, giving rise to greasiness; by the simultaneous use of other plasticisers the greasiness may be diminished. The admixture of an equal part of tricresyl phosphate gives good results. It is a solvent for ester gum and colophony. Ethyl and benzyl cellulose can be plasticised with castor oil.

Acetylated castor has also been used as a plasticiser, but without pronounced success.

Blown castor oil is occasionally used as a plasticiser; its characteristics vary according to the conditions of manufacture; in general they are sp. gr. about 0.98; Acetyl value 150–165; iodine number 70; saponification value 180–190. The more prolonged the blowing the higher are the figures with the exception of the iodine number, which falls.

Butylacetyl ricinoleate

This plasticiser has been recommended as a substitute for castor oil. It is a non-volatile light-coloured oil having a mild odour of castor oil; it has a limited plasticising action on cellulose nitrate but none on cellulose acetate; it can be used in quantities up to 100% with the nitrate before causing synaeresis (sweating). It is one of the best plasticisers for polyvinyl chloride required for electrical insulation at low temperatures. A mixture containing 33% of butyl acetyl ricinoleate has elec. cond. 1×10^{-11} and remains flexible at $-70°$ C (brittle temperature $-70°$ C to $-75°$ C) [6]. It improves light and heat stability and resistance to the action of water; it has a lubricating and anti-sticking effect in moulding processes. It also plasticises polyvinyl acetochloride and is useful with cellulose nitrate for improving gloss.

Characteristics: sp. gr. $0.922–0.927_{25}$; n_D 1.457–1.459; flash p. $230°$ F ($110°$ C); m. p. cloudy at $-32°$ C, solid at $-65°$ C; b. p. about $400°$ C; visc. 0.25; dissolves about 2% of water at $25°$ C. It is a solvent for ester gum, manila and crepe rubber. It partly dissolves rosin, cumarone, mastic, dammar but not shellac, sandarac or arabic.

Benzyl abietate. $C_{19}H_{29} . COO . CH_2 . C_6H_5$

Known also as benzyl resinate and as 'Resin Ether' [8] ; is a neutral semi-solid resinous plasticiser similar in appearance to Canada balsam. It is soluble in all the usual solvents. sp. gr. about 1.04; n_D about 1.55. It is practically non-volatile and is a very stable ester. It can replace both resin and plasticiser and is useful for plasticising ester gum, bakelite resin and copal ester or for obtaining a soft but non-sticky film of cellulose nitrate or acetate, the addition of a small proportion of triacetin being sometimes of advantage with the two last mentioned. It is recommended for leather cloth, and for imparting adhesion, good weathering properties and gloss to cellulose nitrate films.

Ethyl abietate. $C_{19}H_{29} . COO . C_2H_5$

This is a light amber-coloured viscous liquid. sp. gr. 1.03; optical rotation about $-6°$ C; n_D 1.52–1.53; flash p. $352°$ F ($178°$ C); ign. p. $216°$ C (open test); m. p. $-34°$ C approx; b. p. about $350°$ C; Soluble in all the usual anhydrous solvents; dissolves dammar, mastic, ester gum; insoluble in water.

Oxidises very slowly to a hard brittle resin, and tends to darken in colour. It is best used in conjunction with other plasticisers; it blends well with castor oil.

Methyl abietate. $C_{19}H_{29}$. COO . CH_3

A pale yellow viscous liquid having latent solvent properties for cellulose nitrate which are developed by admixture with alcohols. It is soluble in all the usual organic solvents with the exception of methanol. It is compatible with ethyl cellulose, rubber chloride, polyvinyl chloride and chloro-acetate, alkyl resins, coumarone, ester gum, colophony, dammar, mastic, elemi, raw rubber, urea-formaldehyde resin, carnauba and asphalt. It is partly compatible with cellulose aceto propionate and acetobutyrate, paraffin wax, copal, kauri, sandarac, shellac, manila, pontianac but not with cellulose acetate or polyvinyl acetate.

It tends to oxidise and discolour on exposure but it imparts body, gloss and water resistance to cellulose nitrate and ethyl cellulose.

Characteristics [7] : sp. gr. $1.020-1.030_{20}$; n_D $1.5295-1.5300$; b. p. $360°-365°$ C; m. p. below $-40°$ C; flash p. $374°$ F $(180°$ C); ign. p. $218°$ C; sp. ht. 0.395; visc. $28-34$ at $25°$ C; spf. rot. $+13°$ C; dielec. const. 3.65 at $25°$ C, 3.23 at $100°$ C; power factor 0.2% at $25°$ C and 60 cyc., 43.5% at $100°$ C; esters $92-94\%$; acid value 6 max.

Methyl dihydroabietate $C_{19}H_{31}$. COO . CH_3

Obtained by hydrogenating methyl abietate [7], it is less prone to oxidise and discolour but less stable to heat.

Its solvent characteristics are almost identical with those of methyl abietate.

It has n_D $1.517-1.519$; spf. rat. $+36°$ C; dielec. const. 3.5 at $25°$ C, 3.14 at $100°$ C; power factor 0.9% at $25°$ C, 20.5% at $100°$ C.

p-Toluene sulphonamide. p-CH_3 . C_6H_4 . SO_2 . NH_2

Known also as p-toluene sulphamide. It is a white odourless solid m. p. $136-7°$ C, soluble in acetone, ethyl butyland amyl alcohols, and their acetates, glycol ethers, benzene and most organic solvents with the exception of paraffin hydrocarbons. Alcohol dissolves 7.5% at $5°$ C. Water dissolves 0.3% at $20°$ C. Various other sulphamides have been proposed for use as plasticisers, such as the mono- and dimethyl and ethyl sulphonamides of benzene, toluene and xylene [2], and mixtures, some of which are liquids. *Methyl p-toluene sulphamide*, m. p. 78° C, has properties similar to p-toluene sulphamide. All these substances plasticise cellulose acetate, cellulose acetobutyrate, cellulose nitrate, ethyl cellulose, melamine formaldehyde, shellac, vinyl resins and zein.

p-Toluene sulphanilide $p\text{-}CH_3 . C_6H_4 . SO_2 . C_6H_4 . NH_2$

A white crystalline solid m. p. 103° C, easily soluble inctone, alcohol, benzene, ethyl acetate and most organic solvents. It can be used to soften cellulose acetate in proportions up to 50% by weight of the cellulose acetate; it yields rather brittle films. *Methyl p-toluene sulphanilide* is a similar substance, m. p. 95° C.

Ethyl acetanilide $C_6H_5N . (C_2H_5) . CO . CH_3$

Known industrially as 'Mannol,' has a stabilising action on cellulose nitrate, which it gelatinises in a manner similar to that of camphor. In the past it has enjoyed great popularity as a substitute for camphor, its odour being but faint. It is a white crystalline solid. sp. gr. 1.087; m. p. 54° C; b. p. 258° C; flash p. 255° F (124° C); soluble in most organic solvents (70% in alcohol), and slightly in water.

Acetanilide is sometimes used for similar purposes. It is a white odourless crystalline solid. m. p. 115° C; b. p. 304° C; sp. gr. at 20° C 1.20; flash p. 329° F (165° C).

sym-diethyl diphenyl urea
$$\begin{array}{c} C_2H_5 \qquad\qquad C_2H_5 \\ \diagdown\qquad\qquad\diagup \\ N.CO.N \\ \diagup\qquad\qquad\diagdown \\ C_6H_5 \qquad\qquad C_6H_5 \end{array}$$

This substance is a stabiliser as well as a softener for cellulose nitrate. It is a white crystalline solid having a slight peppery odour. m. p. 72°–73° C; b. p. 325°–330° C; sp. gr. 1.12 at 20° C; flash p. 302° F (150° C). Insoluble in water, but soluble in most organic solvents. It plasticises shellac.

Methylphthalylethyl glycollate [5]
$$\begin{array}{c} COO.CH_3 \\ COO.CH_2.COO.C_2H_5 \end{array}$$

A nearly colourless liquid plasticiser for cellulose nitrate and acetate, ehtyl and benzyl cellulose, polyvinyl chloride, polystyrene, rubber chloride and some alkyl resins. It is miscible with most organic solvents but not with the paraffins, castor, linseed or chinawood oils. It is stable to light. sp. gr. 1.215–1.225 at 25° C; m. p. about −35° C; flash p.

375° F $(190^\circ$ C$)$; n_D $1.502-1.506$ at 25° C; visc. 1671_0, 455_{10}, 100_{25}; water dissolves 0.09% at 30° C, 0.05% at 20° C.

Ethylphthalylethyl glycollate [5]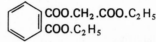

A nearly colourless light-stable plasticiser for cellulose nitrate and acetate, perbunan, formvar and other vinyl resins. It is miscible with most organic solvents except the petroleums; it is partly miscible with vegetable oils. sp. gr. $1.175-1.185_{25}$; m. p. 20° C; flash p. 380° F $(193^\circ$ C$)$; n_D $1.496-1.500$ at 25° C; visc. 611_0, 216_{10}, 63_{25} water dissolves 0.018% at 30° C.

Butylphthalylbutyl glycollate

$$\begin{array}{l} COO.CH_2.COO.C_4H_9 \\ COO.C_4H_9 \end{array}$$

A nearly colourless liquid plasticiser for cellulose nitrate and high acetyl cellulose acetate and acetobutyrate, rubber chloride and polystyrene. It is miscible with all the usual organic solvents and vegetable oils. sp. gr. $1.092-1.102_{25}$; flash p. 380° F $(193^\circ$ C$)$; n_D $1.488-1.492$ at 25° C; b. p. $340\ell-350^\circ$ C (decomp.); liquid at -45° C; visc. 295_0, 136.5_{10}, 51_{25}; water dissolves 0.001% at 30° C; stable to light and to heat at temperatures up to about 290° C.

Butyl laevulinate $CH_3 . CO . CH_2 . COO . C_4H_9$

Known also as butyl laevulate, is a colourless liquid having a mild odour resembling celery. It is a plasticiser for both cellulose acetate and nitrate, and can be used in proportions up to 100%. It is miscible with the usual organic solvents with the exception of glycols and glycerol. It dissolves 2.1% of water and water dissolves 0.93%. It is a solvent for cellulose esters, ester gum, colophony, coumarone, but not for shellac, dammar or pontianac. sp. gr. 0.9735_{20}; n_D 1.430; b. p. about 238° C; dil. rat. toluene 3.0.

Polyglyceryl acetate

Known industrially as glyacol, this consists mainly of diglyceryl ether tetraacetate, but is usually a highly complex mixture. It is a colourless liquid similar to triacetin in character, but more viscous and possessing a higher boiling-point. b. p. about 340° C; sp. gr. 1.146; n_D 1.4435; dil. rat. toluene 0.9. It is a good plasticiser for cellulose acetate and is used in the manufacture of sheets. Maximum proportion 100%. It plasticises rubber chloride but is not miscible with vegetable oils, it does not attack polythene.

Chloroparaffins

Products obtained by the controlled chlorination of paraffin wax and known as Cereclors [10] are used as plasticisers for polyvinyl chloride, rubber chloride and synthetic rubbers. They are nearly colourless viscous oils of very low volatility, miscible with most oils and anhydrous solvents but not with industrial alcohol or water. Two grades are available:

	1	2
chlorine content	47%	40–42%
viscosity	2.3	2.3
sp. gr. at 25° C	1.23	1.16
flash p. over	390° F	390° F

Methyl naphthalene [14]

Is a straw coloured, mobile, aromatic liquid which is a solvent for some alkyl resins and rubber chloride. The industrial product which is a mixture of mainly mono-methyl naphthalenes has sp. gr. 1.000–1.015; b. r. 230°–260° C ... 80%; flash p. 180° F (82° C); evap. range 1400; pure 1-methyl naphthalene has sp. gr. $1.102_{20\ 4}$; b. p. 244.5° C; m. p. −31° C; n_D 1.6174; 2-methyl naphthalene has b. p. 241° C; m. p. 34.2° C.

Amyl naphthalenes

The amyl naphthalenes are light brown viscous oils which are used as plasticisers or extenders for ethyl cellulose, rubber chloride, natural rubber, polystyrene and other resins, particularly when required for high frequency insulators [17] with power factors in the region of 0.001 at 10^6 cycles.

Pure 1-amyl naphthalene has sp. gr. $0.966_{20\ 4}$; b. p. 307° C; m. p. −26° C; n_D 1.5728; 2-methyl naphthalene has sp. gr. $0.956_{20\ 4}$; b. p. 310° C; m. p. −4°; n_D 1.5694.

They are stable to heat, probably non-toxic, non-volatile and are miscible with all the usual solvents but not with methanol or water.

The industrial products [11] have:

	Mono-beta-amyl	Diamyl
sp. gr. at 20° C	0.96–0.97	0.93–0.94
b. r. (° C)	279–330	329–336
flash p. (° F)	255	315
n_D	1.573	1.554
cub. exp.	0.00077	0.00077
lat. ht.	64.3	50.6
sp. ht.	0.411	0.417
visc.	10.1_{20}	89_{25}

Pine Oils

The term 'Pine Oil' is taken to denote the natural essential oils derived from pine trees such as Pinus palustris and not containing the relatively low-boiling terpenes such as pinene, limonene, terpinene or terpinolene. They consist largely of terpineol (50–70%), to which their solvent properties are mainly due, together with borneol (5–10%), fenchyl alcohol (5–10%), camphor, menthane, anethole, methyl chavicole, cineole, dihydro terpineol and sesqui terpenes.

These oils are colourless, or nearly so, have somewhat powerful but pleasant odours and good solvent properties for ethyl cellulose, kauri, colophony, ester gum, coumarone and some albertols, glyptals and bakelites. The products which are available industrially vary considerably in characteristics; the following figures represent the extreme limits; sp. gr. 0.925–0.945. n_D 1.475–1.485. b. r. 180°–250° C. flash p. about 170° F (77° C).

Well-known industrial products of this type are Yarmor pine oil and Herco pine oil [7].

A typical composition of Yarmor pine oil is:

Alpha-terpineol	58.9%
Terpinenol 4	8.3%
Borneol and alpha-fenchyl alcohol	7.8%
Isoborneol, dihydro-alpha-terpineol and beta-fenchyl alcohol	10.7%
Anethole and methyl chavicole	5.3%
Ketones	9.0%

Polyvinyl acetate

Vinyl acetate, in the polymerised form, is a solid resin which has the property of plasticising cellulose nitrate, with which it is miscible in all proportions without producing synaeresis; it is not miscible with cellulose acetate.

Polymerised vinyl acetate is available in several grades, varying in softening point from about $100°$ C to $170°$ C, the viscosity of solutions of these grades vary considerably with the solvent. The refractive index of all forms is about 1.47 and the specific gravity about 1.19 at $20°$ C; they are stable to heat up to $200°$ C, and are soluble in all the usual organic solvents with the exception of the paraffin hydrocarbons, glycols, ether and carbon disulphide.

References

[1] Howards & Sons Ltd., Ilford, London, E.
[2] U.S. Pat. 1740854.
[3] International Union of Chemistry.
[4] Am. Chem. Abs., 1942, p. 6243.
[5] Monsanto Chemicals Ltd., Victoria Street. London, S.W.1.
[6] Courtesy of Imperial Chemical Industries Ltd.
[7] Hercules Powder Co. Inc., Wilmington, Delaware, U.S.A.
[8] A. Boake, Roberts & Co. Ltd., London, E.15.
[9] Cf Brit. Pat. 567705. British Industrial Plastics.
[10] Imperial Chemical Industries Ltd., London.
[11] Sharples Chemicals Inc. Aires Export Corp., N.Y., U.S.A.
[12] B.P. 435, 272.
[13] The Geigy Co., Ltd., Manchester 17.
[14] Shell Chemicals Ltd., London, W.1.
[15] Albright and Wilson Ltd., London, W.1.
[16] Union Carbide Ltd. London
[17] U.S.P. 2,414,497.
[18] Hunter, Perry and Evans. Brit. J. Indust. Med., 1944, I, 227; Chem. and Ind., 1954, p. 674.
[19] Distillers Co. London.

Appendix I: Trade names

Abalyn	Methyl abietate.
Abracol 203	Paratoluolsulfanilid.
Abracol 243	Dicresyl glyceryl ether.
Abracol 777	Mono-cresyl ether diacetate.
Abracol 789	Paratoluolsulfamid.
Abracol 888	Dicresyl glyceryl ether acetate.
Abracol 1001	Tertiary butyl phenol.
Abracol 1011	Mono-cresyl glyceryl ether.
Abwaschmittel N 61	Methylacetate, ethyl propionate and methanol, mixture.
Acetonal, schwer	Higher aliphatic ketones.
Acetonal, weiss	Methylethyl ketone and lower homologues.
Aceton-Ersatz	Various low-boiling ester mixtures.
Acetin	Acetyl glycerin.
Acetosal	Tetrachlorethane.
Actylol	Ethyl lactate.
Acytol	Ethyl lactate.
Adinol	Triethyl citrate.
Adipol BCA	Dibutyl cellosolve adipate.
Adipol 2 EH	Diethyl hexyl adipate.
ADM 100	Linseed oil.
ADM 150	Soya bean oil.
Adronol	Cyclohexanol.
Adronol acetate	Cyclohexanyl acetate.
Agfa fixator	Benzyl salicylate.
Alanol	Tetrachlorethane.
Alcoltate	Petroleum residual denaturant.
Aldehol	Petroleum residual denaturant.
Alexipon	Ethylacetyl salicylate.
Alkyl glycol	Monomethyl and butyl glycol ethers.
Alloprene	Chlorinated rubber.
Alphanol 79	Nonyl alcohol mixture.
Altal	Triphenyl phosphate.
Anol	Cyclohexanol.
Anon	Cyclohexanone.
Anozol (Anazol)	Diethyl phthalate.
Ansol	Ethyl alcohol and ethyl acetate.
Ansol M	Anhydrous ethyl alcohol, denatured.
Ansol ML	Ethyl alcohol 85%, ethyl acetate 6%, butanol 3% and benzol 6%.
Ansol PR	Ethyl acetate substitute containing alcohol.
Antodyne	Mono-phenyl ester.
Apco thinner	Mineral spirit.
Apex No. 1	Dibutoxyethyl phthalate.
Apex No. 2	Diethyleneglycol propionate.
Apex No. 3	Dibutyl ethoxy succinate.
Apex No. 4	Butyl stearate.
Apex No. 5	Lauryl butoxy ethoxy acetate.
Apex No. 6	Dimethyl ethoxy phthalate.
Arklone P	1,1,2-trichlorotrifluoroethane.
Arklone L	1,1,2-trichlorotrifluoroethane/isopropanol azeotrope

Arklone E	1,1,2-trichlorotrifluoroethane/methylene chloride azeotrope
Arklone W	1,1,2-trichlorotrifluoroethane/water surfactant blend.
Arneel TOD	Octadecadine nitrile.
Arneel S	Octadecadine nitrile.
Arneel DN	Dimer of above.
Arneel HF	Fatty acid nitriles.
Arochlor	Polychlorodiphenyl.
Asordin	Carbon tetrachloride.
Avantine	Isopropyl alcohol.
B2	An alkyl ricinoleate.
B3 reagent	Terpene tertiary alcohols.
Baker's No. 1	Methyl ricinoleate.
Baker's No. 2	Ethyl ricinoleate.
Baker's P3	Butyl ricinoleate.
Baker's P4	Methyl acetyl ricinoleate.
Baker's P.4.C	Methyl cellosolve acetyl ricinoleate.
Baker's P5	Ethyl acetyl ricinoleate.
Baker's P6	Butyl acetyl ricinoleate.
Baker's P7	Methyl undecylinate.
Baker's P8	Acetylated castor oil.
Baker's P9	Acetylated castor oil, polymerised.
Baker's P11	Methyl ester of polymerised ricinoleic acid.
Baker's P12	Ethyl ester of polymerised ricinoleic acid.
Baker's P13	Butyl ester of polymerised ricinoleic acid.
Baker's P14	P11 acetylated.
Baker's P15	P12 acetylated.
Baker's P16	P13 acetylated.
Banana oil	Amyl acetate.
Bardol	Coal tar oil.
Barkite	Methyl cyclohexanol oxalate.
Barkite B	Dimethyl cyclohexanol oxalate.
Beckolak	Benzene and ethyl alcohol.
Benzalin	Nitrobenzol.
Benzinoform	Carbon tetrachloride.
Benzinol	Trichlorethylene.
Benzolinar	Ethyl ether 1 and benzol 4.
Benzosol	Guaiacol benzoate.
Benzsuccin	Dibenzyl succinate.
BGC	Butoxyethyl diglycol carbonate.
Bisoflex DBS	Dibutyl sebacate.
Bisoflex DNA	Adipate of 3,5,5-trimethylhexanol.
Bisoflex DNS	Sebacate of 3,5,5-trimethylhexanol.
Bisoflex DOA	Di(2-ethylhexyl) adipate.
Bisoflex DOS	Di(2-ethylhexyl) sebacate.
Bisoflex 79 Adipate	Adipic acid ester $C_7 - C_9$ alcohols.
Bisoflex 79 Sebacate	Sebacic acid ester $C_7 - C_9$ alcohols.
Bisoflex 81	Di(2-ethylhexyl) phthalate.
Bisoflex 88	Phthalic acid ester C_8 alcohols.
Bisoflex 91	Phthalic acid ester nonyl alcohol.
Bisoflex 100	Diisodecyl phthalate.
Bisoflex 100 Adipate	Diisodecyl adipate.
Bisoflex 102	Dicaprylate ester of triethylene glycol.

Bisoflex 103	Multiple ester.
Bisoflex 108	Phthalate ester of mixed alcohols.
Bisoflex 108 Adipate	Adipate ester of mixed alcohols.
Bisoflex 791	Phthalate ester $C_7 - C_9$ alcohols.
Blacosolv	Trichlorethylene.
Bonoform	Tetrachlorethane.
Butacol	Butyl lactate.
Butanone	Methyl ethyl ketone.
Butol	Butyl butyrate.
Butoxyl	Methyl butylene glycol acetate.
Butyl carbitol	Diethylene glycol mono butyl ether.
Butyl cellosolve	Ethylene glycol mono butyl ether.
Butyl dioxitol	Diethylene glycol mono butyl ether
Butyl dioxitol acetate	Diethylene glycol mono butyl ether acetate
Butyl glykol	Ethylene glycol mono butyl ether.
Butyl oxitol	Ethylene glycol mono butyl ether.
Byk special	Methyl acetate techn.
C24	Methyl amyl di-hexyl cyclohexanone.
Calcitone	Acetone and methyl acetate.
Calol ethacate	Denaturant.
Camphol	Oxanilide.
Camphrosal	p-Toluene sulphonamide.
Canadol	Light benzine.
Carbinol	Mixture of terpenes.
Carbitol	Diethylene glycol mono-ethyl ether.
Carbona	Carbon tetrachloride.
Casterol	Esterified castor oil.
Cecoline No. 1	Trichlorethylene.
Cecoline No. 2	Tetrachlorethylene.
Cellon	Tetrachlorethane.
Cellosolve	Ethylene glycol monoethyl ether.
Celludol	Toluene sulphonamide.
Celluflex M179	Tricresyl phosphate.
Celluflex M142	Aromatic phosphate.
Cellusol	p-Toluene sulphonamide.
Centralite 1	Diethyl diphenyl urea.
Centralite 2	Dimethyl diphenyl urea.
Centralite 3	Methylethyldiphenyl urea.
Centralite 4	Ethylphenyl ethyl o-toluidine urea.
Cereclor	Chlorinated paraffin wax.
Cetamoll	Mixture of phosphoric esters.
Cetamoll QU	Chlorethyl phosphate.
Chlorasol	Carbon tetrachloride.
Chlorex	BB Dichloroethyl ether.
Chlorylen	Trichlorethylene.
Circosolve	Trichlorethylene.
Colcine	A plasticiser and partially fixed oil.
Comedol	Trichlorethylene.
Columbus spirit	Methyl alcohol.
Clophen A60	Chlorinated diphenyl.
Cresylin	Cresyl glyceryl ether.
Crexylol	Toluene substitute.
Cyclohexal acetate	Cyclohexanyl acetate.

Cyclonol	Methyl cyclohexanone glyceryl acetal.
DCP	Dicapryl phthalate.
Dec	Decahydronaphthalene.
Decanap	Decahydronaphthalene.
Dekalin	Dekalin, decahydronaphthaline.
Depanol	A turpentine product.
Depol	A turpentine product.
Desmodurs	Polyglycol isocyanates.
Di48	Dichlorethylene.
Di60	Dichlorethylene.
Diatol	Diethyl carbonate 90%.
Diacetin	Glyceryl diacetate.
Diaphanol	Methylcyclohexanyl acetate.
Diathylin	Glyceryl diacetate.
DICE	Dichlorodiethyl ether.
Dichlorditane	p-dichlorodiphenyl methane.
Dicresylin	Dicresyl glyceryl ether.
Dieline	Dichlorethylene.
Diluols	Paraldehyde mixtures.
Diluol 3	Anhydrous ethyanol and petroleum.
Diluent D	Mineral spirit.
Dinopol	Di-n-octyl phthalate.
Dional	Ethyl acetate mixture.
Dioxitol	Diethylene glycol mono ethyl ether.
Dipentene 122	Dipentene 50% and other terpenes.
Dissolvan	Methyl and ethyl acetate, alcohols and acetols.
Dissolvan CA	Low boiling ester mixture.
Dissolvan DN	Low boiling acetal mixture.
DNOP	Di-n-octyl phthalate.
DOP	Di-2-ethylhexyl phthalate.
DOS	Diethylhexyl sebacate.
Dow P1	Diphenyl t-butylphenyl phosphate.
Dow P2	Mono-phenyl di-t-butylphenyl phosphate.
Dow PP3	Diphenyl-o-chlor phenyl phosphate.
Dow P4	Monophenyl dichlorophenyl phosphate.
Dow P5	Diphenyl-o-xylenyl phosphate.
Dow P6	Monophenyl dixylenyl phosphate.
Dow P7	Tri-p-t-butyl phenyl phosphate.
Drawin 28	Mainly ethyl acetate.
Drawin Ch	Trichlor ethylene (stabilised).
Duatol	Guaiacol acetate.
Dutch liquid	Ethylene dichloride.
Dynalcol	1 alcohol and 1 benzol.
Eclenol 344	Dimethyl cyclohexyl phthalate.
Elaol	Dibutyl phthalate.
Elaol I	Hexanetriol hexoate.
Elaol II	Hexanetriol octoate.
Elastol	p-Toluene ethyl sulphonamide.
EMA	Methyl acetate and acetoacetic ester.
Emaillet	Tetrachlorethane.
Enrodin	Glycol chlorhydrin mixture.
Erganol	Dibenzyl ether.
Ergol	Benzyl benzoate.

Escon	Ethyl acetyl salicylate.
Essigester	Ethyl acetate.
Ester P	Ethyl propionate.
Estergemische 13	Low-boiling ester mixture.
Esterol	Dibenzyl succinate.
Estisol	Ethyl lactate.
Eteline	Perchloroethylene.
Ethox	Ethoxy ethyl phthalate.
Ethylene diacetate	Ethylene glycol diacetate.
Ethyl lactol	Petroleum fraction and alcohol.
Eufin	Diethyl carbonate.
Eufixin	Isoamyl benzoyl glycollate.
Eusolvan	Ethyl lactate.
Exalgen	Methyl acetanilide.
Exluan	Dioxane.
Estersols	Ester mixtures of various boiling ranges.
Fermine	Dimethyl phthalate.
Ferrissol	Guaiacyl cinnamate.
Flek-flip	Trichlorethylene.
Flexalyn	Diethylene glycol diabietate.
Flexalyn C	Ethylene glycol diabietate.
Flexol DOP	Di-2-ethylhexyl phthalate.
Flexol 2 GCP	Dichlorethoxyethyl phthalate.
Flexol 3GH	Triethylene glycol di-2-ethyl butyrate.
Flexol 3GO	Triethylene glycol di-2-ethyl hexoate.
Flexol 4GO	Polyethyleneglycol di-2-ethyl hexoate.
Flexol CS24	Di-2-ethyl cellosolve succinate.
Flexol 3CF	Trichloroethyl phosphate.
Flexol 3MP	Ditrimethylcyclohexyl phthalate.
Flexol TOF	Tri-2-ethylhexyl phosphate.
Flexol 8HP	Diethylhexyl tetrachydrophthalate.
Flexol DHP	Di-n-hexyl phthalate.
Flexol A26	Diethylhexyl adipate.
Flexol TWS	A Thioester.
Flexol 8N8	Diethylhexamidodiethyl hexoate.
Flexol B400	A polyalkyl glycol derivative.
Flexol R2H	A polyester.
Flexol CC55	Diethylhexyl hexahydrophthalate.
Flexol 2GB	Diglycol dibenzoate.
Formosol	Ethyl formate.
GB ester	Butyl glycollate.
G18	Glycidyl Oleate.
Genklene P	1,1,1-trichloroethane.
Genklene	1,1,1-trichloroethane (stabilised).
Genklene VA	Genklene blend of low volatility.
Gensol	Turpentine.
Glyakol	Diglyceryl ether tetra acetate.
Glycine A	Diethanol sulphide.
Glycoester	Glycolmonoacetat.
Glykosal	Glyceryl salicylate.
Grundlage	Dimethyl phthalate.
Haftex	Chlorinated naphthalene.
Halowax	Chlorinated naphthalene.

Halowax 4001 B2	Chlorinated hydrocarbons.
Heptalin	Methyl cyclohexanol.
Heptanaphthene	Methyl cyclohexane.
Hercolyn	Hydrogenated methyl abietate.
Hercosol 5	Pine oil derivative.
Hercosol 80	Terpene hydrocarbons and ketones.
Hexachlorene	Hexachlorethane.
Hexahydrokresol	Methyl cyclohexanol.
Hexalin	Cyclohexanol.
Hexalin acetate	Cyclohexanyl acetate.
Hexogen	Hexone.
Hexone	Methylisobutyl ketone.
Hexanon	Cyclohexanol.
Hydropalat A	Diethyl hydrophthalate.
Hydropalat B	Dibutyl hydrophthalate.
Hydroterpin	Hydrogenated turpentine product.
Hexeton	Methyl isopropyl cyclohexanon.
Hexyl ketol	Ketone mixture B.R. $150^{\circ}-170^{\circ}$
Hydremil	Amyl nitrite.
Hydrolinol	Methyl cyclohexanol and spirit.
Hydrolin	Methyl cyclohexanol.
Imsol A	Isopropanol-water azeotrope.
Imsol M	Isopropanol-water azeotrope with methanol.
Industrial spirit	Ethyl alcohol.
Isopropyglykol	Ethylenglycol monoisopropylether.
Isopral	Trichloroisopropyl alcohol.
Isopryl	Methylisopropyl ether.
Itrosyl	Ethyl nitrite.
KP23	Butyl cellosolve stearate.
KP45	Diethyleneglycol di-propionate.
KP61	Mixed aliphatic and phthalic esters.
KP77	Dimethyl Thianthrene.
KP120	Methyl cellosolve acetyl ricinoleate.
KP140	Tributyl cellosolve phosphate.
KP150	Dilanryl phthalate.
Kalosche	Benzine.
Kapsol	Methyl cellosolve oleate.
Katarine	Carbon tetrachloride.
Kelalin	Decahydronaphthalene.
Kemsolene	Mineral spirit.
Kessol MA	MEK and anhydrous alcohol.
Ketonone B	Butyl benzoylbenzoate.
Ketonone E	Ethyl o-benzoylbenzoate.
Ketonone M	Methyl o-benzoylbenzoate.
Ketonone MO	Methoxy ethyl benzoylbenzoate.
Ketol	Ketone mixture, B.R. 60°-200°
Kronilyne	Tri-phenyl phosphate.
Kronisol	Di-butoxyethyl phthalate.
Kronitex AA	Tri-cresyl phosphate.
Krystallol	Petroleum distillate.
Lactole spirit	Petroleum distillate, B.R. 80°-130°.
Lactonal	Ethyl acetate.
Lak	Amyl acetate and benzol.

Lembitol	Ethyl lactate.
Lewisol	Ester gum.
Ligroin	Low-boiling petroleum.
Limonene	Dipentene.
Lindol	Tri-cresyl phosphate.
Losungsmittel LC	Ethyl acetate substitute containing acetals.
Losungsmittel APV	Polyglycolether product.
Losungsmittel BP	Butyl propionate.
Losungsmittel C	Acetal.
Losungsmittel DA	Diacetone alcohol.
Losungsmittel E13	Mixture of methyl acetate, ethylacetate and methanol.
Losungsmittel E14	Mixture of methyl acetate, ethylacetate and methanol.
Losungsmittel E33	Mixture of low-boiling esters, etc.
Losungsmittel EMA	Mixture of low-boiling esters, etc.
Losungsmittel G	Glycol.
Losungsmittel GA	Ethylglycol.
Losungsmittel GAC	Ethylglycol acetate.
Losungsmittel GC	Glycol monoacetate.
Losungsmittel GD	Glycol diacetate.
Losungsmittel GM	Methylglycol.
Losungsmittel GMC	Methylglycol acetate.
Losungsmittel GO	Glycol diacetate.
Losungsmittel L30	Terpenes.
Losungsmittel O	Dibutyl oxalate.
Losungsmittel RS200	Butyrolactone and valerolactone.
Losungsmittel TAD	Mixed chlorohydrocarbons.
Losungsmittel THD	Dichlor ethylene.
Losungsmittel LWH	Methylene dichloride.
Losungsmittel 8951	Isopropyl alcohol.
M.142	Aromatic phosphate.
M.179	Aromatic phosphate.
M.180	Isopropyl o-cresyl phosphate.
Mannol	Ethyl acetanilide.
MC50	Methyl alcohol, methyl acetate and ethyl acetate.
MC60	Methyl alcohol, methyl acetate and ethyl acetate.
MC75	Methyl alcohol, methyl acetate and ethyl acetate.
MEK	Methyl ethyl ketone.
Mesamolls	Alkyl-sulphonic aryl esters.
Metacelludol	m-Toluene sulphamide.
Metal	Mainly low-boiling esters.
Metopryl	Methyl n-propyl ether.
Methox	Methoxy-ethyl phthalate.
Methylaceton	Acetone, methyl acetate, methanol.
Methlanol	Methylcyclohexanol.
Methone	Methyl acetate and acetone.
Methyl adronal	Methylcyclohexanol.
Methyl adronal acetate	Methyl cyclohexanyl acetate.
Methyl anol	Methyl cyclohexanol.
Methyl anon	Methyl cyclohexanone.
Methyl cellosolve	Ethylene glycol monomethyl ether.
Methyl cellosolve acetate	Ethylene glycol monomethyl ether acetate.
Methyl glykol	Ethylene glycol monomethyl ether.
Methyl glykol acetat	Ethylene glycol monomethyl acetate.

Methyl hexalin	Methyl cyclohexanol.
Methyl hexalin acetate	Methyl cyclohexanol acetate.
Methylhexanon	Methyl cyclohexanone.
Methyl oxitol	Ethylene glycol mono methyl ether.
Methyl oxitol acetate	Ethylene glycol mono methyl ether acetate.
Methynol	Methyl alcohol.
Mineral spirit No.1	Petroleum distillate, b.r. $150°$-$215°$.
Mineral spirit No.2	Petroleum distillate, b.r. $80°$-$130°$.
Mittel AEP	Ethyl toluene sulphonate.
Mittel ADT	Acetanilide.
Mittel B3	A sulphonamide.
Mittel B4	A sulphonamide.
Mittel B5	A sulphonamide.
Mittel B6	Alkyl toluene sulphonamide.
Mittel B7	A sulphonamide.
Mittel HNA	Acetophenone.
Mittel KP	p-cresyl toluene sulphonate.
Mittel OC	Dibutyl oxalate.
Mittel L30	Product from turpentine.
Mittel PA	Cellosolve phthalate.
Mittel PA	Ethyl glycol phthalate.
Mittel PH	Hydroxy naphthanilide.
Mittel PI	p-Toluene sulphanilide.
Mittel PJ	Paratoluol sulphanilide.
Mittel PM	Dimethyl glycol phthalate.
Mittel PO	p-Toluene phenyl sulphonate.
Mittel P11	2-3-Oxynaphtholic acid-anilide.
Mollit A	Glyceryl acetyl benzoate.
Mollit B	Glyceryl tribenzoate.
Mollit BR extra	Glycerine ester mixture.
Mollit I	Diethyldiphenyl urea.
Mollit II	Dimethyldiphenyl urea.
Mollit IV	Diethylotolyl urea.
Monoil	Xylene methyl-sulphonamide.
Monoplex 7	Butoxyethyl sebacate.
Monoplex 16	A nitrile.
Morpholine	Diethyl ether imide.
Naftolen	High boiling hydrocarbon.
Naftolen MV	Lubricating oil extract.
Naphthene	Cyclohexane.
Naphtol AIS	Hydroxynaphthanilide.
Natalit	3 Spirit, 2 ether.
Neantin	Diethyl phthalate.
Neoline	Dinonyl naphthalene.
Neu-camphrosal	p-Toluene sulphamide.
Nevsol	A toluene substitute.
Nilox ester oil	Methyl cyclohexanol oleate.
Niobe oil	Methyl benzoate.
Normanol	Ethyl lactate.
Novania	Tetrachlorethane.
Octoil	2-Ethyl hexyl phthalate.
Oktalin	Octohydronaphthalene.
Opal wax	Hydrogenated castor oil.

Organosal	Resin in plasticiser and thinner.
Orthene	o-Dichlorbenzene.
Oxitol	Ethylene glycol mono ethyl ether.
Oxitol acetate	Ethylene glycol mono ethyl ether acetate.
Ozonin	Ozonised turpentine oil.
P3	Ethylbutyl ricinoleate.
P4	Methyacetyl ricinoleate.
PC4	Methyl cellosolve acetyl ricinoleate.
P5	Ethylacetyl ricinoleate.
P6	Butylacetyl ricinoleate.
P8	Acetylated castor oil.
P9	Acetylated polymerised castor oil.
P11	Methyl ester of polymerised ricinoleic acid.
P14	P11 acetylated.
P16	Butyl acetyl polyricinoleate.
P40	Dichloro diphenyl sulphone.
P135	Di-methylamide of aliphatic acids.
P1C	Methyl cellosolve ricinoleate.
P2C	Cellosolve ricinoleate.
P6C	Butyl cellosolve acetyl ricinoleate.
P6E	Ethylbutyl ricinoleate.
Palatinol A	Diethyl phthalate.
Palatinol BB	Benzylbutyl phthalate.
Palatinol C	Di-n-butyl phthalate.
Palatinol E	Diethyl glycol phthalate
Palatinol JC	Di-isobutyl phthalate.
Palatinol M	Dimethyl phthalate.
Palatinol O	Dimethyl glycol phthalate.
Paracetat	Isopropyl acetate.
Paraplex	Glycol sebacate.
Paraplex G25	Sebacic acid polyester.
P.C.125	Ethyl acetate and butyl propionate.
Pentacetate	Synthetic amyl acetate.
Pentalene	Amyl naphthalenes.
Pentalin	Pentachlorethane.
Pentalyn	Pentaerythritol abietate.
Pentaryl A	Monoamyl diphenyl.
Pentaryl B	Diamyl diphenyl.
Pentasol	Synthetic amyl alcohol.
Pentasol acetate	Synthetic amyl acetate.
Pentexel	Synthetic amyl acetate and amyl alcohol.
Peracol	Cyclohexanol and ethyl alcohol.
Peramyl alcohol	Cyclohexanol, butyl alcohol and isopropyl alcohol.
Perawin	Perchlorethylene.
Perklone	Perchloroethylene for dry cleaning.
Perchloroethylene Grade 1	Perchloroethylene.
Perma	Chlorinated naphthalene.
Perma-A-chor	Trichlorethylene.
Permetol	Methyl cyclohexanol and tetrachlorethane.
Permitol	Chlorinated diphenyl.
Petrobenzol	Mineral spirit.
Persprit	Isopropyl alcohol.
Perudin	Benzyl benzoate.

Peruscabin	Benzyl benzoate.
Petrohol	Isopropyl alcohol.
PG16	Acetylated butyl polyricinoleate.
Phoenixine	Carbon tetrachloride.
Phonepine	Carbon tetrachloride.
Pinene III	α-pinene 92% and other terpenes.
Pinolin	Rosin spirit.
Pinradol	Pine root oil.
Placidol A	Diamyl phthalate.
Placidol B	Dibutyl phthalate.
Placidol E	Diethyl phthalate.
Plassitil	Glyceryl monolactate triacetate.
Plastic A	Glyceryl tri-benzoate.
Plastic X	Tricresyl phosphate.
Plasticiser 30	Butyl capryl phthalate.
Plasticiser 64	Hexachlordiphenyl.
Plasticiser 101	Dimethyl glycol phthalate.
Plasticiser SC	Triglycol dioctylate.
Plasticiser 50B	Butyl cyclohexyl phthalate.
Plastisols	Resin in plasticiser.
Plastoform 1	Benzyl alcohol.
Plastoform 11	Phenyl ethyl alcohol.
Plastol A	Toluene sulphonamide.
Plastol C II	Mixed aromatic sulphonamides, mainly toluene sulphonamide.
Plastol M	Alkylated toluene sulphonamide mixture.
Plastol V.G.	Ethyl toluene sulphonamide.
Plastol V.B.	
Plastolein 9050	Dihexyl azelate.
Plastolein 9055	Diethylene glycol dipelargonate.
Plastolin I	Benzyl abietate.
Plastolin II	Amyl salicylate.
Plastomol P	Toluene sulphoanilide and toluene methyl sulphonamide.
Plastoplex	Butyl phosphate.
Pleck	Di-butyl phthalate.
Polane	Chlormethyl stearate.
Polysolvan AN	Higher aliphatic acetates.
Polysolvan HSN	Ester of aliphatic alcohols.
Polysolvan SNSN	Ester of higher aliphatic alcohols.
Polysolvan O	Ester of glycolic acid.
Polysolvan HS	Methyl cyclohexanone and acetate.
Prestone	Diethylene glycol.
Pumilin	Pine root oil.
Puran	Monochlorbenzene.
PY 3E	Ethy butyl soy -acid esters.
PY 16	Acetylated butyl polyricinoleate.
Pyranton	Spirit, acetone and diketone or cyclohexanol and cyclohexanone.
Pyranton A	Diacetone alcohol.
Pyrene	Carbon tetrachloride.
Quittnerlack	Tetrachlorethane.
R.B.A.	Butyl acetyl ricinoleate.
R.E.A.	Ethyl acetyl ricinoleate.

Reomol E	Alkyl glycol phosphate.
Reomol G	An aryl glycol ether.
Reomol J	Alkyl glycol phosphate.
Reomol NF1	Trichlor-phenyl phosphate.
Reomol NF2	Trichlor-ethyl phosphate.
Reomol P	Methyl cellosolve phthalate.
Reomol DCP	Dicapryl phthalate.
Reomol DHP	Dihexyl phthalate.
Reomol DOP	Diethylhexyl phthalate.
Reomol DOS	Diethylhexyl sebacate.
Reomol DP	Diphenyl phthalate.
Renol	Alcohol and carbon tetrachloride.
Resilvestrol	Synthetic balsam turpentine oil.
Resin ether	Benzyl abietate.
Ricol	Blown castor oil.
Rixolin	Petroleum and camphor oil.
Rodol N 333	Methyl alcohol and methyl acetate.
S 454	Glycol oxo-pentanoate.
S 747	An ether-alcohol phosphate.
S 787	Methyl cellosolve stearate.
Salacetol	Ethyl acetyl salicylate.
Salicyline	Glyceryl mono-salicylate.
Salvarvol	Di-ethyl phthalate.
Sangajol	White spirit, b,p. about 160° C.
Santiciser 1	p-Toluene sulphonamide.
Santiciser 2	Ethyl p-toluene sulphonamide.
Santiciser 3	p-Toluene N-ethyl sulphonamide.
Santiciser 7	m-Toluene ethyl sulphonamide.
Santiciser 8	o- and p-Toluene N-ethyl sulphonamides.
Santiciser 9	o-p-Toluene sulphonamides.
Santiciser 10	o-Cresyl p-toluene sulphonate.
Santiciser E15	Ethyl phthalyl ethyl glycollate.
Santiciser B16	Butyl phthalyl butyl glycollate.
Santiciser H	Cyclohexyl p-toluene sulphonamide.
Santiciser M10	o-p-m-Cresyl benzene sulphonate.
Santiciser M17	Methyl phthalyl ethyl glycollate.
Santiciser M81	Methyl o-p-toluene sulphonamide.
Santiciser 127	N-n-butyl-benzene sulphonamide.
Santiciser 128	N-ethyl benzene sulphonamide.
Santiciser 130	N-isopropyl benzene sulphonamide.
Santiciser 131	130 with toluene homologue.
Santiciser 139	p-Toluene sulphonamide.
Santiciser 140	Monocresyl diphenyl phosphate.
Santiciser 160	Butyl benzyl phthalate.
S.C.	Pine tar.
S.C. Plasticiser	Tri-ethylene glycol laurate.
Sectol	Methyl cyclohexanol.
Sepalin M.O.J.	Isopropyl methyl adipate.
Sericosol	A glycol derivative.
Sextate	Methyl cyclohexanyl acetate.
Sextol	Methyl cyclohexanol.
Sextone	Cyclohexanone.
Sextone B	Methyl cyclohexanone.

Shellacol	Alcohol petroleum mixture.
Sinclair solvents	High naphthene paraffins.
Sipalin AOC	Dicyclohexanyl adipate.
Sipalin AOM	Dimethylcyclohexanyl adipate.
Sipalin MOM	Dimethylcyclohexanyl methyl-adipate.
Sipalin Special	Sipalin MOM and a palmitic ester.
Skellysolve	Naphtha fraction.
Skellysolve PA	Di-ethyl glycol phthalate.
Skellysolve PM	Di-methyl glycol phthalate.
Skellysolve REA	Acetyl ethyl ricinoleate.
SOA	Sucrose octa-acetate.
Softener 9	A tribasic alcohol.
Softener 77	Thio dibutyl acetate.
Softener 88	Methylene dithiobutyl acetate.
Softener 90	Hydroxyaliphatic ester of tribasic alcohol.
Softener 99	Thio diethylene cyanide.
Solactol	Ethyl lactate.
Solasthin	Methylene dichloride.
Solazetol	Ethyl acetyl salicylate.
Solisol	Isopropyl alcohol.
Solox	100 denatured alcohol, 5 ethyl acetate and 1 gasoline.
Solvaloid C	Aromatic hydrocarbon.
Solvantine	Methyl isobutyl carbinol.
Solvarol.	Diethyl phthalate.
Solvarone	Dimethyl phthalate.
Solvatone	Acetone 80%, isopropyl alcohol 10% and toluene 10%.
Solvenol No. 1	Dipentene and other terpenes from pine oil.
Solvene	Coal tar solvent naphtha, b.r. 160°-190° C.
Solvent GC	Ethylene glycol mono-acetate.
Solvent PE	Butyl acetate ex paraffins.
Solvent E8	Mixture of methanol and a volatile acetate.
Solvent E10	Mixture of methanol, a volatile acetate and a volatile ketone
Solvent E12	Mixture of ethyl acetate, methyl acetate, methanol and butanol.
Solvent E13	Mixture of higher volatile acetates methanol and butanol.
Solvent E15	Mixture of volatile acetates and methanol.
Solvent E16	Mixture of ethyl acetate and medium boilers including butanol.
Solvent E33	Methyl acetate and alcohol.
Solvent E82	Mixture of ethyl and butyl acetates.
Solveol	Diethyl phthalate.
Solvesso	Hydrogenated naphtha.
Solvesso 1	Toluol petroleum fraction.
Solvesso 2	Xylol petroleum fraction.
Solvohol	Glycol and polyglycol ethers.
Solvulose	Ethylene glycol mono ethyl ether.
Spasmodin	Benzyl benzoate.
Spasmin	Dibenzyl succinate.
Spectrol	Carbon tetrachloride.
Spezial-losungsmittel	Methyl acetate and methanol.
A	Methyl acetate, methanol and acetone.
AE and EF	Mainly low-boiling esters.

E	Methyl acetate, ethyl acetate ad methanol.
Hiag	Methyl acetate and methanol.
Verein	Methyl acetate, acetone and methanol.
EE33	Methyl acetate, ethyl acetate and methanol.
EF, E13	Methyl acetate, ethyl acetate and methanol.
AE14	Methyl acetate, acetone and methanol.
Spirits of wine	Ethyl alcohol.
Spiritogen	Methyl alcohol.
Spirosal	Glycol mono salicylate.
Spritol	Methyl alcohol.
Stabilisal A	Nitrogenous polymerisation product.
Staybelite 2	Diethylene glycol ester of hydrogenated rosin.
Staybelite 3	Modified Staybelite 2.
Stryacol	Guaiacol cinnamate.
Sunco spirits	Petroleum fraction b.r. 150°-215° C.
Sumatrin	Sumatra petrolbenzin.
Suresnol	Diethyl phthalate.
Sylon	An amino-silane.
Synthin	Synthetic petroleum hydrocarbon.
Tamasol	Butyl acetate.
Tamasol J	Isobutyl acetate.
Tamasol II	Higher aliphatic alcohols with cyclic ketones.
Tamasol III	Higher aliphatic alcohols with cyclic ketones.
Tamatol JN	n-Butyl acetate.
T.C.P.	Tricresyl phosphate.
Tegin	Glyceryl monostearate.
Tegofan	Chlorinated rubber.
Tekol	Purified linseed stand-oil.
Terapin	Petroleum fraction.
Terlitol	Petroleum fraction.
Terposol 3	Terpinyl methyl ethers.
Terposol 8	Terpinyl glycol ethers.
Terpozone	Ozonised turpentine.
Tetra	Carbon tetrachloride.
Tetracol	Carbon tetrachloride.
Tetraform	Carbon tetrachloride.
Tetralex	Perchlorethylene.
Tetralin	Tetrahydro naphthalene.
Tetranap	Tetrahydro naphthalene.
Textile	Petroleum fraction, b.r. 65°-100° C.
Theolin	American wood turpentine.
Thiokol TP-90B	A polyether formal.
TNP	Trinaphthyl phosphate.
TOF	Triethylhexyl phosphate.
Tollac	A toluene substitute.
Tolulene	Mineral spirit.
Tornext	Chlorinated rubber.
TPP	Triphenyl phosphate.
Tri	Trichlorethylene.
Triklone A	Trichloroethylene (amine stabilised)
Triklone N	Trichloroethylene (neutral stabilised).
Tribenzoin	Glyceryl tribenzoate.
Tributyrin	Glyceryl tributyrate.

Tricarbin	Clyceryl tricarbonate.
Trieline	Trichloroethylene.
Troluoil	Petroleum fraction, b.r. 90°-130° C.
TTXP	Tritolylxylenyl phosphate.
Turmintine	Petroleum fraction, b.r. 150°-215° C.
TXP	Trixylenyl phosphate.
Ulol	A mixture of esters.
Utanol	Sulphurated bitumen.
Vamolin	High-boiling petroleum.
Vansol	50 Toluene and VMP naphtha.
Varnolene	Petroleum fraction, b.r. 150°-210° C.
Varsol	Mineral spirit, Sp. gr. 0.782, flash p. 40° F. (5°)
VMP naphtha	Mineral spirit, b.r. 100°-160° C.
Vulcanol B	High-boiling aromatic hydrocarbons.
Westron	Tetrachloroethane (obsolete name)
Westrol	Trichloroethylene (obsolete name)
Wood spirit	Crude methyl alcohol.
Wood naphtha	Crude methyl alcohol.
Wood alcohol	Crude methyl alcohol.
Xylidin	Nitroxylene.
Yarmor pine oil	Terpineol and terpene alcohols.

Appendix II: Solubility tables

APPENDIX II

SOLUBILITY TABLES

The solubilities quoted in these tables must be regarded as merely indicative and should be interpreted in the following senses: :

S = Soluble to an extent deemed to be sufficient for technical purposes, the solutions obtained being such that clarification by settling, filtering or centrifuging is practicable without excessive waste.

P = Partly, moderately or difficultly soluble.

N = Not sufficiently soluble for practical purposes.

X = Soluble under certain conditions which are mentioned in the text.

There is an inevitable element of uncertainty in tables such as these on account of the variable nature of the substances under consideration. In general, it may be taken that the solubilities quoted for the natural resins refer to a good average unadulterated quality containing natural wax. The solubilities for cellulose nitrate refer to ½ sec NC, those for cellulose acetate refer mainly to the B.E.S.A. 2D50 quality. (53–53·5% acetyl as acetic acid) and those for ethyl cellulose ranging 47–48% ethoxyl.

Solvents.	Cellulose nitrate	Cellulose acetate	Ethyl cellulose	Benzyl cellulose.	Rubber	Rubber chloride.	Ester gum.	Polystyrene.	Polyvinylacetate	Polyvinylchloride	Polyvinylchloroacetate	Shellac.	Hard copal	Soft manila	Hard manila	Dammar.	Kauri	Mastic	Sandarac	Elemi	Congo	Cumarone	Colophony	Paraffin hydrocarbon	Vegetable oil	Castor oil	Methyl methacrylate
HYDROCARBONS																											
Benzene	N	N	S	X	S	N	S	S	S			N	N	P	N	P	P	S	N	S	P	S	S	S	S	S	S
Toluene	N	N	S	X	S	S	S	S	S	N	P	N	N	P	N	P	P	S	N	S	N	S	S	S	S	S	S
Xylene	X	N	S	S	S	S	S	S	P	N	P	N	N	P	N	S	P	S	N	S	N	S	S	S	S	S	S
Tetrahydronaphthalene	N	N	S		S	S		S	N				N			N	N	S				S	S		S		
Decahydronaphthalene	N	N	S		S								N			S	N	S					S		S		
Dipentene (Limonene)	N	N	P		S	S	S									S	S					S	S				
Turpentine	N	N	P				S		N							P										N	

VM and P Naphtha, Petroleum	N	N	N	N	S	N	S	N	N	N	N	N	N	N	N				N							N	
Nitropropane	S	N	S	S	N	S	S		S	P	S	N	N	S	P	P	P	S			N	S	P	N	S	S	
ALCOHOLS																											
Methyl alcohol	X	N	S		N	N	N	N	S	N	N	S	N	P	N	P	P	P	P	N	P	N	S	N	N	S	N
Ethyl alcohol	N	N	S	X	N	N	N	N	S	N	N	S	N	S		S	S	P	S	S	P	X	S	X	S	S	N
n-Propyl alcohol	N	N	X	X	N	N	S	N	P	N	N	S	N	S		S	S	P	S	S		N	S	S	S	S	
Isopropyl alcohol	N	N	X		N	N	N	N	P	N	N	S	N	S	P	S	S	P	S	S	S	N	S			S	P
n-Butyl alcohol	N	N	X	N	N	N	S	N	P	N	N	S	S	S	P	S	S	P	S	S	S	P	S	S	S	S	P
Isobutyl alcohol	N	N	X		N	N	S	N	P	N	N	S		S		S	S	P	S	S	S	N	S	S	S	S	
sec-Butyl alcohol	N	N	N		N	S	S	N		N	N	S		S		S	S	P	S	S	S		S		S	S	
Amyl alcohol	N	N	P		N		S	N	P	N	N	S		S		N	S	P	S	S	S	P	S	S	S	S	
Benzyl alcohol	P	S	S	S	N	N	S	S	S	N		S	S	S	S	S	S	S	S	S		S	S	N	S	S	
Diacetone alcohol	S	S	P	P	N	N	P	N	S	N	P	S	S	S	S	N	S	S	S	P	S	S	S	N	S	S	
Diethyl ether	X	N	N	N	P		S	S	N	N	N	N		S	P	P	N		S	S	N		S		S		N
Diisopropyl ether	X	N	P	N	S		S		N	N	N	N				P	N						S	S	S		
KETONES																											
Acetone	S	S	S	X	X	S	S	S	S	P	S	P	P	S	S	P	P	P	S	S	P	S	S	S	S	S	P
Methyl ethyl ketone	S	X	S			S	S		S			P		S		P	S		S		S	S	S	S	S	S	
Methyl propyl ketone	S	N	S			S	S		S			P		S		P	S					S	S				
Methyl amyl ketone	S	N	S			S	S		S	S	S	N		P		P	S	S	S	S		S	S	S	S	S	
Methyl isobutyl ketone	S	X	S	S	S	S	S		S	S	S	N		P		S	P	S	S	S		S	S	S	S	S	
Butyrone; di-n-propyl ketone	P	P	S			S	S		S					S								S	S	S	S	S	P

Material	Methyl acetate	Ethyl acetate	n-Propyl acetate	iso-propyl acetate	n-Butyl acetate	sec-Butyl acetate	Isobutyl acetate	Amyl acetate	Hexyl acetate	n-Butyl formate	Amyl formate	n-Butyl proprionate	Amyl propionate	Ethyl butyrate
Methyl methacrylate	N			P		P	P							
Castor oil	S	S	S	S	S	S	S	S	S	S	S	S	S	
Vegetable oil	S	S	S	S	S	S	S	S	S	S	S	S	S	
Paraffin hydrocarbon	S	S	S	S	S	S	S	S	S	S	S	S	S	
Colophony	S	S	S	S	S	S	S	S	S	S	S	S	S	S
Cumarone	N	S	S	S	S	S	S	S	S	S	S	S	S	S
Congo	N	N			S		S					N	N	
Elemi	N	N	X	S	S	S	S	S	S	S	S	S		
Sandarac	S	S	P	S	S	S		S	S	S		N	N	S
Mastic	P	S	S	S	S	S	S	S	S	S	S	S	S	S
Kauri	S	S	P	S	S	S	S	S	S			S	S	
Dammar	N	N	X	N	P	P	P	P	S	S		N	N	N
Hard manila			P			S								
Soft manila	S	S	P	S	P	S	S	S	S	S		N	S	S
Hard copal	N	N	N	N	N	N	N	N			N	N	N	N
Shellac	N	P	P	P	X	P	P	P	N	P	P	P	N	
Polyvinylchloroacetate		S		S	S			⊗	S					
Polyvinylchloride	N	X		P	S			S	S					P
Polyvinylacetate	S	S	S	S	S		S	S	S			S	S	
Polystyrene	S	S	S	S	S	S	S	S	S			S		S
Ester gum	S	S	S	S	S	S	S	S	S	S	S	S	S	S
Rubber chloride	N	S		S	S	S		S				S		
Rubber	N	N		N			N	N		S				
Benzyl cellulose														
Ethyl cellulose	X	S	S	S	S	S	S							S
Cellulose acetate	S	S	X	N	N	N	N	N	N	N	N	N	N	N
Cellulose nitrate	S	S	S	S	S	S	S	S	S	S	S	S		S

ESTERS

Solvent																							
n-Butyl butyrate	S	S	S	S	S	S		S	S	S		N		S						S				N
Benzyl formate	S	S	S	S	S	S		S	S			N	N							S			N	
Benzyl acetate	S	S	S	X	S	S		S	S	P		N	N	N	X	S		S	S	N		N	N	
Ethyl lactate	S	S		S	S			S	S	S		P	S	S	S	S		S	S	S	P	S	S	S
n-Butyl lactate	S	S	S	S	S	S		S	S	S		N	N		P	P	S	S	S	S	P	S	N	P	S
Amyl lactate			X	X		X			X	N			N	P		N	S		P	N	S	P	N	S	S
Butyl glycollate	S				S				X	X		S		S	S	S	P	P	S			P	S
Methyl benzoate	N					S	S		S	X		N	N	S	S	N	P	S	S	P	P		N	S
Ethyl benzoate	S		S	S	P	S		S	S			P	S	S	P	S	S	S	S	P	P	P	P	S
Diethyl carbonate			S	S	P	N				X		N	N	N	S	S	S	S	S		N	N	P	S
Diethyl carbonate	S		P	P	S					X		N	N	N	P	S	S	N	S	P	P	N	N		P
GLYCOLS						N	X								P	S		P	S	S			N	
Ethylene glycol	S		S	P	P	S			X	X					S	S		S	S		P		N	
Ethylene glycol monoacetate	N	X	X		X			X	N	N			N	N		X	N	N	N			N	N	S
Ethylene glycol diacetate	S		X	S	P	S	S		S	N		P	N	N	S	P	S	S	S		P	X	N	S
Ethylene glycol monomethyl ether					N					N			S	N	N	S			N			N	N	N
Ethylene glycol monoethyl ether		P	S	S	P	S			S			N	N	N	S	S	N	N	S		S	N		S
Ethylene glycol diethyl ether			S	S	P								S	S		N	N	S				S		
Ethylene glycol ethyl ether acetate	S	S	S	N	S	S	S	S	S			N	N	N	P	P	S	S	P	N	S	N	Z	P
Ethylene glycol monobutyl ether			N	N	P		P		N				N		Z	P	S	S				N		
Butylene glycol		Z		N	S								Z		N		S	S	P		Z	N		
Butylene glycol diacetate			S		S		X		Z	Z			Z		Z	S	S	N	S			Z		
Diethylene glycol	N	S	S	S	P	Z	S	S	X	N		N	S	P	S	X	N	S	S	N	S	N	N	N
Diethylene glycol monoacetate	S	S	S	S		P		P	X					S	Z	S	Z	S	S		S	S	S	
Diethylene glycol monoethyl ether	S	S	S	S	S	P	N	S	X		X	S	S	S	Z	N	S	Z	S	N	S	S	S	S

Solubility of resins in solvents (S = soluble, P = partly soluble, B = swells, N = insoluble, X = special, SP = swells/partly soluble):

Resin	Dioxane	Propylene glycol	CYCLOHEXANES	Cyclohexanol	Cyclohexanol acetate	Methyl cyclohexanol	Methyl cyclohexanol acetate	Cyclohexanone	Methyl cyclohexanone	CHLORO-COMPOUNDS	Mono-chlorhydrin	Dichlorhydrin	Epichlorhydrin	Methylene dichloride	Chloroform	Carbon tetrachloride
Methyl methacrylate	S			P	S			S	S					S	S	N
Castor oil											S					
Vegetable oil		N		S		S					N	S		S	S	S
Paraffin hydrocarbon		N		S	S	S	B	S	S		N	N				
Colophony		S		S	S	S	S	S	S						S	S
Cumarone	S			S		S		S	S		N	S	S			S
Congo				P	P	P	P	B	P		N				P	P
Elemi	S			P	S	P	S	B	S			S	S		S	S
Sandarac	S			N	S	N		S	S			S	P		P	P
Mastic	S			S	S	S	S	S	S		P	S	S		S	S
Kauri	S	P		S	S	S	S	S	S		N	S	P	P	P	P
Dammar	S	N		P	S	P	S	B	P			S	P	P	P	S
Hard manila				S	S	S	S	SP	S		N		N			N
Soft manila	S			S	S	S	S	S	S			S	P		P	P
Hard copal				P	P	P	P	S	S		N		N		N	N
Shellac	S	P		S	S	S	P	S	S		P	S	P		N	N
Polyvinylchloroacetate	S	N						S	S					S	N	N
Polyvinylchloride	S	N						S	S					S	N	N
Polyvinylacetate	S	N		P	S	P		S	S					S	P	S
	S			N	S	N		S	S					S	S	S
Ester gum	S	N		S	S	S	S	S	S		P	S	S			S
Rubber chloride				P	P	S		S	S						S	S
Rubber	S	N		P	S			S	S					S	S	S
				S	S	S		S	S							
Ethyl cellulose	S			S	S	S		S	S		S	P	S		S	S
Cellulose acetate	S	N		N	P	N	P	S	S		S	S	S	S	X	X
Cellulose nitrate	X	N		N	S	N	S	S	S		N	X	S	S		N

Material																							
Dichloroethane	S	S	S	S	S	N	P	P	P	N	S	P	N	N	N	S	S	S	S	S	S	X		S
Tetrachloroethane		S	S	S	S	N		S	S	N	S	S		N	N	S	S	S	S	S	S	X		S
Perchloroethane			S	S	S	N		N	S	N		N	N	N	N	S		S	S	S	S	S		S
Dichloroethylene	S	N	S	S	S		S	N	S	N	N	N		N			S	S	S	S	N	S		S
Trichloroethylene		N	S	S	S		S	N		N	N	N	N	N	N	N		N	S	S	S	N	X	X
Perchloroethylene				S	S	S	S			N	N	N	N	N	N	N		N	S	S	S	N	S	S
Monochlorobenzene	S	S	S	S		S	S	N	S	S		N	N	N	N	S			S	S	S	N	N	N
Dichloroethyl ether	S	S	N	S	S	S		S		N	N	N		N	N		P	S	S	S	N	X	X	N
1,1,2-trichlorotrifluoroethane	P	S	S		S	N		N	N		N	N	N	N	N		N	N	N	N	P	N	N	N

FURFURALS

Material																								
Furfural						S	N	S	S		N		S	S	S									S
Furfuryl alcohol						P	N	N	N		S	S	P	P	X	S						S	X	S
Tetrahydrofurfuryl alcohol						N		S	S		S	S	S	S			S		S	S	S	S	S	S

PLASTICISERS

Material																								
Acetophenone	S	S	S		S						N	S	S	S	S	S		X	S	S	N	S	X	S
Diacetin	P	N	N	N	N	N					P	S		S	N	P		N	S	N	N	S	S	S
Triacetin	N	N	N	P	P	P					N	S		S	N	S		N	S	P	N	S	N	S
Triphenyl phosphate	S	S	S		S						X	S	S		S	S		X	S	S	S	S	N	S
Tricresyl phosphate	S	S	S		S						N	S	X		S	S	S	X	S	S	N	S	N	S
Butyl oleate	S	S	S								N	S	N					S	N	N	P	S	N	N
Butyl stearate	S	S	S								N	N	N					N	N	N	P	S	N	N
Amyl stearate	S	S	S								P	P	P					N	N	N	P	N		N
Amyl benzoate	S	S	S	P							X	P	P					P	P	P	P	P		P
Benzyl benzoate	S	S	S	S	S	S					S	X	X	S	S	S		P	S	S	S	P	S	S

Material	Dimethyl phthalate	Diethyl phthalate	Dibutyl phthalate	Diamyl phthalate	Dihexyl phthalate	Dioctyl phthalate	Dibutyl sebacate	Dioctyl sebacate	Triethyl citrate	Tributyl citrate	Triamyl citrate	Dibutyl tartrate	Diamyl tartrate	Cresyl glyceryl diacetate	Methyl abietate
Methyl methacrylate		S	S	S	S							S			
Castor oil	S	S	S	S								S	S		
Vegetable oil	X	S	S	S	S	S	S	S		S		P	S	S	
Paraffin hydrocarbon	N	P	S	S	S	S			S	S	S	S	S	S	
Colophony	S	S	S									S	S	S	S
Cumarone	S	S	S	S	S	S			S	S	S	S			S
Congo															P
Elemi														S	P
Sandarac															P
Mastic	P	S	S									S	P	S	
Kauri													N	P	
Dammar	P	X	S	S	S	S			S	S	S	S	S	S	
Hard manila	X		X											P	P
Soft manila													P	P	
Hard copal	X	X	X									N	N		
Shellac	P	P	P	P	P	P	N	S	S	S	S	S	S		P
Polyvinylchloroacetate	S	S	S	S	S	S	S	S	S	S		S	S		S
Polyvinylchloride	N	X	S	S	S	S	S	S	X	S	S	N	P	S	S
Polyvinylacetate	S	S	S	S	S	N		N	S	S	S	S	S		N
Polystyrene	S	S	S	S	S	S				P	N	P	N		
Ester gum	S	S	S	S	P	S			S	S	S	S	S	S	S
Rubber chloride	S	S	S	S			S	S			S	S			
Rubber	S		P			S									S
Benzyl cellulose		S												S	
Ethyl cellulose	S	S	S	S	S	S	S	S		S	S	X	X	S	S
Cellulose acetate	S	S	X	X	X	N	N	N	S	P	N	S	S	S	N
Cellulose nitrate	S	S	S	S	S	S	S	S	S	S	S	S	S	S	X

Appendix III: Plasticiser proportions

Plasticiser	Percentage proportions beyond which cloudy, oily or excessively soft films result with	
	Cellulose nitrate	Cellulose acetate
Diamyl phthalate	100	10
Amyl stearate	2	5
Diamyl tartrate	100	80
Benzyl alcohol	150	75
Benzyl benzoate	100	66
Tribenzyl citrate	75	20
Benzyl laevulinate	50	90
Butyl acetyl ricinoleate	100	–
Tributyl citrate	75	20
Butyl laevulinate	100	100
Butyl oleate	2	5
Dibutyl phthalate	100	10
Butyl stearate	2	5
Dibutyl tartrate	150	100
Butylene glycol diacetate	100	100
Butylene glycol dibenzoate	75	30
Butylene glycol monolactate	85	85
Butylene glycol oxalate	75	75
Dibutyl glycol phthalate	–	33
Castor oil	3	–
Cresyl glyceryl diacetate	100	100
Cyclohexyl laevulinate	100	100
Triethyl citrate	100	100
Diethyl-glycol phthalate	100	50
Diethyl phthalate	80	100
Glyceryl tribenzoate	150	20
Glycol dilaevulinate	25	60
Dihexyl phthalate	100	15
Dilauryl phthalate	66	66
Methyl cyclohexyl oxalate	150	66
Dimethyl-glycol phthalate	50	50
Dimethyl phthalate	75	80
Diphenyl phthalate	80	50
Triacetin	66	66
Trichlorphenyl phosphate	75	50
Tricresyl phosphate	100	10
Triphenyl phosphate	100	10
Vinyl acetate	∞	0

Index

abbreviations, 87
Abel apparatus, 65
acetal, 134
acetone, 75, 136-8
acetophenone, 208
acetylated castor oil, 231
acetylene dichloride, 193
acetylene tetrachloride, 74, 192-3
adhesion of lacquer films, 22, 49
ageing, resistance to, 18
alcohol, *see* ethyl alcohol
alcohols, 111-31; heats of combustion of, 59; molecular association in, 53
American Conference of Governmental Industrial Hygienists, 71
American Society for Testing Materials, 86
amyl-, *see also* diamyl-, triamyl-
n-amyl acetate, 26, 151-3
sec-amyl acetate, 153-4
amyl acetates, 123-4
amyl alcohols, 122-8
amyl benzoate, 21, 216
amyl chlorides, 123-4, 200-1
amylene glycols, 206
amyl formate, 77, 156
amyl ketone, 141
amyl lactate, 162
amyl naphthalene, 236-7
amyl oxalate, 23
amyl propionate, 157
amyl stearate, 21, 215-16
aromatic coefficient, 14
auto-ignition temperatures, 60-1
azeotropic mixtures, 27-8, 32, 43-4, 45-6, 54

barrier creams, 70
benzene, 38, 71-2, 88-90; legal requirements about, 63, 66, 89
Benzene Solvents Regulations, 89
benzine, 50
benzyl abietate, 232
benzyl acetate, 159
benzyl alcohol, 28, 68, 130
benzyl benzoate, 217
benzyl butyl phthalate, 224
benzyl formate, 158
blush, 21, 26, 28, 46, 50; resistance to, 17, 31, 47
blush numbers, 47

body, 49
boiling points of solvents: and flash point, 56; and explosivity, 59; and solvent power, 27; and vapour pressure, 42-3; and viscosity, 36, 37-8
British Pharmacopaeia, 101
British Railways regulations, 55-6, 68-9
British Standards Institution, 38, 86
brittleness of lacquer films, 16, 17, 19, 38
brushing properties, 37, 38
butanol, *see* butyl alcohol
butyl-, *see also* dibutyl-, tributyl-
n-butyl acetate, 28, 78, 149-50
sec-butyl acetate, 150
butyl acetyl ricinoleate, 232
n-butyl alcohol, 28, 79, 117-20
sec-butyl alcohol, 35, 121
tert-butyl alcohol, 122
butyl benzene, 14
butyl benzoate, 216
n-butyl butyrate, 158
sec-butyl carbinol, 122
butyl cellosolve, 76, 171
butylene glycol, 172
butylene glycol diacetate, 173
butylene glycol dibenzoate, 217
butylene glycol ethyl ether, 172
butylene glycol monolactate, 229
butyl formate, 77, 154, 156
butyl furfuryl ether, 206
butyl furoate, 206
n-butyl lactate, 161-2
sec-butyl lactate, 162
butyl laevulinate, 235
butyl oleate, 214
butyl phthalyl butyl glycollate, 235
n-butyl propionate, 157
butyl stearate, 215
p-tert-butyl toluene, 73
butyrone, 141

camphor, 17, 23, 230-1
carbitol, 174
carbon disulphide, 63, 79, 107-8
carbon tetrachloride, 55, 63, 67, 74, 188-9
castor oil, 231-2; as plasticiser, 18, 20, 21, 22, 23, 24
cellosolve, 169

cellosolve phthalate, 223
cellulose lacquers, constituents of, 26
cellulose nitrate: combustion of, 56;
    nitrogen content of, and viscosity
    of solution, 5; polar groups in, 6;
    solutions of, 8-14
Cellulose Solutions Regulations, 65-6,
    68
Cerechlors, 236
chilling, 26, 27, 29, 45, 48-9; resistance
    to, 17, 32
chloro-compounds, 184-202; see also
    dichlor-, trichlor-
2-chloro-ethyl alcohol, 79, 184
chloroform, 67, 73, 187-8
chloro-nitroparaffins, 109-10
chloro-paraffins, 236
coal tar solvent naphtha, 95
cohesive energy density, 6-7
colloidal solutions, 4
constant viscosity procedure, 13
coordinate graphs, 29-33
cordite acetone, 137
cost, calculations of, 31, 33
cotton blush, 21, 26, 28, 46, 50
cresyl glyceryl ethers, 229-30
critical solution temperature, 20-1
cumene, 73
cyclohexane, 98, 100
cyclohexane derivatives, 179-83
cyclohexanol, 28, 43, 179-80
cyclohexanone, 43, 76, 182
cyclohexanyl acetate, 180
cyclohexanyl stearate, 216
cymene, 96

"dangerous goods", transport of, 67,
    69
decahydronaphthalene (decalin), 107
n-decane, 105
dew point, 48
diacetin (glyceryl diacetate), 208-9
diacetone alcohol, 28, 130-1
diamyl phthalate, 220
diamyl tartrate, 228
dibutyl ether, 133
dibutyl glycol phthalate, 223
dibutyl phthalate, 219
dibutyl sebacate, 225-6
dibutyl tartrate, 227-8
dicapryl phthalate, 221
dichlorbenzene, 67
dichlorethane, 74, 189-90

dichlorethylene, 38, 193-4
dichlorethyl ether, 78, 199-200
dichlormethane, 73, 186-7
dichlorpentane, 201
dichlorpropanes, 200
dicresyl glyceryl ether, 229-30
dicresyl glyceryl ether monoacetate,
    230
diethoxy ethyl phthalate, 223
diethyl acetal, 134
diethyl butyl phthalate, 220
diethyl carbonate, 163-4
diethyl diphenyl urea, 234
diethylene dioxide, 176
diethylene glycol 173-4
diethylene glycol dibenzoate, 217
diethylene glycol monoacetate, 174
diethylene glycol monobutyl ether,
    175-6
diethylene glycol monoethyl ether,
    174-5
diethyl ether, see ethyl ether
diethyl glycol phthalate, 223
diethyl hexyl adipate, 225
diethyl hexyl sebacate, 226
diethyl hexyl tetrahydrophthalate,
    224-5
diethyl ketone, 141
diethyl phthalate, 218-19
dihexyl phthalates, 220-1
di-isobutyl ketone, 143
di-isobutyl phthalate, 220
di-isopropyl ether, 132
di-isopropyl ketone, 142
dilauryl phthalate, 222
diluents, 26-7, 102-3; non-solvents as,
    45-6, 51; and viscosity, 37-8
dilution ratios, 8-11, 21-3, 27
dimethyl acetal, 133
dimethyl acetone, 141
dimethyl cyclohexyl oxalate, 229
dimethyl formamide, 164-5
dimethyl glycol phthalate, 223
dimethyl heptanone, 143
dimethyl phthalate, 217-18
di-octyl phthalates, 221-2
di-octyl sebacate, 226
dioxane, 80, 176-7
dipentene, 96
diphenyl phthalate, 224
dipropyl ketone, 141
dipole moments, 6, 7, 13
dynamic viscosity, 38

elastic stretch, 23
elasticity, 16, 22, 37
epichlorhydrin, 186
ester gum, 20, 21, 26, 28
esters, 59, 77-8, 144-65
ethane diol, 166
ethanol, *see* ethyl alcohol
ether, *see* ethyl ether
ethers, 78, 131-4
ethoxyethyl acetate, 171
ethoxyethyl alcohol, 169
ethyl abietate, 232-3
ethyl acetanilide, 234
ethyl acetate, 26, 78, 145-7
ethyl alcohol, 79, 113-14
ethyl amyl ketone, 140-1
ethyl benzene, 14, 72, 94
ethyl benzoate, 162-3
ethyl butanol, 128, 155
ethyl butyrate, 158
ethyl cellosolve, 76
ethylene chloride, 55
ethylene chlorhydrin, 79, 184-5·
ethylene dichloride, 74, 189-90
ethylene glycol, 166-7
ethylene glycol diacetate, 167
ethylene glycol diethyl ether, 171
ethylene glycol monoacetate, 167
ethylene glycol monobutyl ether
    (butyl cellosolve), 76-7, 171-2
ethylene glycol monoethyl ether
    (cellosolve), 28, 169-71
ethylene glycol monoisopropyl ether,
    176
ethylene glycol monomethyl ether
    (methyl cellosolve), 76, 167-8
ethylene glycol monomethyl ether
    acetate, 168
ethyl ether, 78, 131-2
ethyl formate, 77, 154
ethyl furoate, 206
ethyl hexanol, 129
ethyl lactate, 28, 159-61
ethyl palmitate, 215
ethyl phthalyl ethyl glycollate, 235
evaporation of solvents, 16-18; effects
    of rate of, 26, 38; latent heat of,
    46-7; tables of rates of, 50, 51, 52;
    vapour pressure and rates of, 41-54;
    viscosity changes during, 36-7
evaporometers, 52
explosive limits, 56-8
explosivity, 45, 57, 59

Factories Act (1961), 63, 74
fire risk, 56
flash point, 45, 64, 65-6; boiling point
    and, 55-6; in legal requirements, 64,
    65-6, 67, 68
flow, 17, 21, 33, 37; viscosity as
    resistance to, 34, 37
fungicidal plasticisers, 24
furaldehyde, 203
furan, 206-7
furanes, 203-7
furfural, 80, 203-4
furfuryl acetate, 205
furfuryl alcohol, 205
furoic (pyromucic) acid, esters of, 206
furyl carbinol, 205
fusel oil, 122, 126; acetates of, 152

gelling of solutions, 5, 8, 20
gloss, 17, 21, 22, 23, 38, 49; secondary
    flow and, 37
glucoside links, hydrolysis of, 36
glyceryl acetates, *see* diacetin, triacetin
glyceryl cresyl ethers, 229-30
glyceryl dichlorhydrin, 185-6
glyceryl monolactate triacetate, 209-10
glyceryl phthalate resins, solvents for,
    12
glyceryl tribenzoate, 210
glycol, *see* ethylene glycol
glycol chlorhydrin, 184
glycols and their ethers, 76-7, 166-78
gum blush, 26, 28, 46, 50

hardness of lacquer films, 49
haze, 28
heats of combustion, 59
heavy oil: definition, 68
n-heptane, 105
heptanol, 129
heptanone, 141
hexahydrocresol, 180
hexahydromethyl-phenol, 180
hexahydro-phenol, 179
n-hexane, 105
hexanols, 127, 128, 155
hexanone, 142
hexyl acetates, 154, 155
high aromatic petroleum
    hydrocarbons, 104
high-boiling-point solvents, 16-17, 28,
    37, 38, 42, 46, 49, 55
high-flash lacquers, 66

hot-spraying of lacquers, 49
hydrocarbon oil: definition, 67-8
hydrocarbons, 6, 88-110; explosive
    limits of, 57; toxicity of, (aromatic)
    71-3, (chlorinated aliphatic) 73-5
hydrogenated solvent naphtha, 100
hydrogen bonding, 7, 13
hydroterpin, 106
hydroxyl groups: and evaporation rate,
    42-3; and polarity of solvents, 6
hydroxymethyl pentanone, 130
hygroscopic and non-hygroscopic
    solvents, 48

ingition temperature, 56
industrial alcohol, 114
inflammability, 55-61; in legal
    terminology, 62-3, 65-6
iso-amyl furoate, 206
isobutanol, see isobutyl alcohol
isobutyl acetate, 150-1
isobutyl alcohol, 120
isobutyl carbinol, 122
isobutyl lactate, 162
isomers, plasticiser properties of, 24
isophorone, 76, 183
isopropyl acetate, 148-9
isopropyl alcohol, 115-17
isopropyl carbinol, 120
isopropyl ether, 78
isopropyl lactate, 161

kauri gum, 28
ketones, 75-6, 136-44
kinematic viscosity, 38

lacquers (cellulose), constituents of, 26
latent heat: of evaporation, 46-7; of
    vaporisation, 6-7
ligroins, 101-2
low-boiling-point solvents, 16, 37, 42,
    46, 55

maximum allowable concentrations, 71
medium-boiling-point solvents, 37, 38,
    42, 46, 55
mesityl oxide, 76, 130, 143-4
methanol, see methyl alcohol
methoxyethanol, 167
methyl-, see also dimethyl-, trimethyl-
methyl abietate, 233
methyl acetate, 77, 145
"methyl acetone", 138-9

methyl alcohol, 78-9, 111-13
methyl n-amyl ketone, 143
methylated spirit, 35, 68, 114
methyl benzoate, 163
methyl bromide, 67
methyl butanols, 122, 124, 125, 126
methyl butyl acetates, 124
methyl n-butyl ketone, 142
methyl carbitol, 175
methyl cellosolve, 76, 167
methyl chloroform, 55, 74, 191
methyl cyclohexane, 100
methyl cyclohexanol, 180-1
methyl cyclohexanone, 183
methyl cyclohexyl acetate, 181
methyl cyclohexyl oxalate, 228-9
methyl cyclohexyl phthalate, 222-3
methyl dihydroabietate, 233
methylene dichloride (dichlormethane),
    55, 67, 73, 186-7
methyl ethyl carbinol, 121
methyl ethylene glycol stearate, 216
methyl ethyl ketone, 75, 139-40
methyl formate, 77
methyl furoate, 206
methyl isobutyl carbinol, 127
methyl isobutyl ketone, 142
methyl isopropyl benzene, 96
methyl lactate, 159
methyl naphthalene, 236
methyl pentanols, 127, 155
methyl pentanone, 142
methyl pentanyl acetates, 155
methyl phenyl ketone, 208
methyl phthalyl ethyl glcollate, 234-5
methyl propyl ketone, 141
methyl p-tolune sulphamide, 234
methyl p-toluene sulphamide, 234
Ministry of Transport rules, 66-9
miscibility of solvents, 7, 8; with
    water, 28-9
mixtures of solvents, 4-6, 35;
    constant-boiling, see azeotropic
    mixtures; evaporation of, 42, 53;
    flash points of, 55; vapour pressures
    of, 43-5
molecular associations, 4-5, 6, 42-3, 53
molecular weight: and evaporation
    rate, 17, 52; of plasticisers 17-18;
    and solvent power, 5; and vapour
    pressure, 46
monochlorbenzene, 199
monochlorhydrin, 185

monocresyl glyceryl ether, 229
monocresyl glyceryl ether diacetate,
   230

naphthas, 94-5; heavy, 14;
   hydrogenated solvent, 100; varnish
   makers' and painters', 105
National Benzole Association, 86
nitrobenzene, 67
nitroethane, 109
nitro groups, and polarity of solvent, 6
nitroparaffins, 74, 108-9
nitropropanes, 79, 109
n-nonane, 105
non-polar solvents, 6, 38
non-solvents: as diluents, 8, 45-6, 51;
   and viscosity, 34-6

n-octane, 105
"orange-peel" effect, 29
OXO process, 129

paraldehyde, 184
Pensky-Marten apparatus, 65
Pentacetate, 153
pentachlorethane, 67, 75, 193
pentane, 105
pentanol acetates, 124
pentanols, 124, 125, 126, 127
pentanones, 141
Pentasol, 122, 125
perchlorethylene, 38, 196-7
perchlormethane, 55
peroxides of ethers, explosive, 132,
   133, 207
petroleum, 63, 100-1
Petroleum Act, 63-5, 66, 68
petroleum ethers, 101-2
petroleum hydrocarbons, 100-5
petroleum mixture, 64
petroleum spirit, 64, 80
phthalates, 23
pimpling, 49
α-pinene, 97
pine oils, 237
pinolin, 98
plasticiser proportions, 261
plasticisers, 15-18, 38; resins as, 18-19;
   retention of solvent by, 19-20; as
   solvents, 8, 50; and viscosity, 37
plasticising solvents, 15-25, 208-38
plasticity, 16, 21
plastics, 24

plastic stretch, 23
poisoning, acute and chronic, 70
polyethylene, "crystalline" polymer, 8
polyglyceryl acetate, 235-6
polymers: plasticisers for, 24;
   solubilities of, 8
polyurethane lacquers, 29
polyvinyl acetate, 237-8
precipitation, 8, 27, 28, 29, 41
pressure, and explosive limits, 57
propanediol, 172
propionone, 141
n-propyl acetate, 43, 78, 147-8
propyl alcohol, 114-15
propylene dichloride, 200
propylene glycol, 172
propyl furoate, 206
PTFE, "crystalline" polymer, 8
pyromucic (furoic) acid, esters of, 206

resins: in cellulose lacquers, 21, 26, 49;
   as plasticisers, 18-19; solvents for,
   12, 14, 28; viscosities of solutions
   of, 38-9
rosin spirit, 98
rubber, solvents for, 102

"sand-papering", 49
saturated solutions, 4
secondary flow, 17, 21, 37
sipalins, 23
solubility, 5
solubility parameters, 7-8, 13
solubility tables, 125, 126, 254-60
solvent action, 4-15
solvent balance, 26-40
solvent power, 5, 8, 27, 35
spraying of lacquers, 29, 35, 49
stabilisers, 132, 191
styrene, 72-3, 105-6
sulphur chloride, 63
synaeresis, 20

T.C.P., 211
temperature: critical solution, 20-1;
   and dilution ratio, 11; and
   explosive limits, 57; in
   lacquer-spraying, 49; lowering of,
   on evaporation, 47-8; and vapour
   pressure, 42
tensile strength, plasticisers and, 21,
   22, 23-4
ternary mixtures, 45

terpenes, 96
tetrachlorethane, 74, 192-3
tetrachlorethylene, 67, 75
tetrachlormethane, *see* carbon
    tetrachloride
tetrahydrofuran, 80, 207
tetrahydrofurfuryl alcohol, 205-6
tetrahydrofurfuryl oleate, 214
tetrahydronaphthalene (tetralin), 73,
    106-7
thinners, 35-6, 105
threshold concentration, dilution ratio
    at, 11
threshold limit values, 71
tolerance, of one solvent for another,
    28
toluene, 38, 72, 90-2; in determination
    of dilution ratios, 9, 10, 11, 12, 22
p-toluene sulphanilide, 234
p-toluene sulphonamide (sulphamide),
    23, 233-4
tolyl xylenyl phosphate, 213
toxicity, 70-81
trade names, 239-52
triacetin, 23, 209
triamyl citrate, 227
triangular coordinate graphs, 29-33
tributyl citrate, 227
tributyl phosphate, 23, 213
1,1,1-trichlorethane, 55, 74, 191
1,1,2-trichlorethane, 190-1
trichlorethylene, 38, 63, 55, 75, 194-5
trichlorethyl phosphate, 213-14
trichlormethane, *see* chloroform
trichlorphenyl phosphate, 214
trichlorfluoroethane, 197-9

tricresyl phosphate, 23, 211-12
triethyl citrate, 226
triethylene glycol monomethyl ether,
    175
triethyl phosphate, 213
trimethyl hexanol, 129
triphenyl phosphate, 56, 210-11; as
    plasticiser, 21, 24
tritolyl phosphate, 52, 211
trixylenyl phosphate, 213
Trouton's formula, 46-7
turpentine, 81, 97-8, 99
"two-type" solvents, 26

vapour pressure: and evaporation rates,
    16, 41-54; and viscosity, 34
vinyl acetate, 237-8
viscometer, falling-sphere, 38
viscosity, 4-5, 33, 34-40; and dilution
    ratio, 13; plasticisers and, 20, 21; of
    plasticisers, 24
volatile solvent, retention and
    elimination of, 17, 19-20, 22, 23
volatility: of plasticisers and resins, 19;
    of solvents, 16-18; and toxicity, 70

water: deposition of, 48 (*see also*
    chilling); solvents miscible with,
    28-9
water blush, 26
white spirit, 14, 80, 103
"wood spirit", 138-9

xylene, 14, 35, 38, 92-4; in
    determination of dilution ratios,
    10, 14